T0181993

IFIP Advances in Information and Communication Technology   **637**

Editor-in-Chief

Kai Rannenberg, Goethe University Frankfurt, Germany

IFIP – The International Federation for Information Processing

IFIP was founded in 1960 under the auspices of UNESCO, following the first World Computer Congress held in Paris the previous year. A federation for societies working in information processing, IFIP's aim is two-fold: to support information processing in the countries of its members and to encourage technology transfer to developing nations. As its mission statement clearly states:

> IFIP is the global non-profit federation of societies of ICT professionals that aims at achieving a worldwide professional and socially responsible development and application of information and communication technologies.

IFIP is a non-profit-making organization, run almost solely by 2500 volunteers. It operates through a number of technical committees and working groups, which organize events and publications. IFIP's events range from large international open conferences to working conferences and local seminars.

The flagship event is the IFIP World Computer Congress, at which both invited and contributed papers are presented. Contributed papers are rigorously refereed and the rejection rate is high.

As with the Congress, participation in the open conferences is open to all and papers may be invited or submitted. Again, submitted papers are stringently refereed.

The working conferences are structured differently. They are usually run by a working group and attendance is generally smaller and occasionally by invitation only. Their purpose is to create an atmosphere conducive to innovation and development. Refereeing is also rigorous and papers are subjected to extensive group discussion.

Publications arising from IFIP events vary. The papers presented at the IFIP World Computer Congress and at open conferences are published as conference proceedings, while the results of the working conferences are often published as collections of selected and edited papers.

IFIP distinguishes three types of institutional membership: Country Representative Members, Members at Large, and Associate Members. The type of organization that can apply for membership is a wide variety and includes national or international societies of individual computer scientists/ICT professionals, associations or federations of such societies, government institutions/government related organizations, national or international research institutes or consortia, universities, academies of sciences, companies, national or international associations or federations of companies.

More information about this series at https://link.springer.com/bookseries/6102

Eunika Mercier-Laurent ·
Gülgün Kayakutlu (Eds.)

Artificial Intelligence for Knowledge Management, Energy, and Sustainability

9th IFIP WG 12.6 and 1st IFIP WG 12.11
International Workshop, AI4KMES 2021
Held at IJCAI 2021
Montreal, QC, Canada, August 19–20, 2021
Revised Selected Papers

Springer

Editors
Eunika Mercier-Laurent (iD)
University of Reims Champagne-Ardenne
Reims, France

Gülgün Kayakutlu (iD)
Istanbul Technical University
Istanbul, Turkey

ISSN 1868-4238   ISSN 1868-422X   (electronic)
IFIP Advances in Information and Communication Technology
ISBN 978-3-030-96594-5   ISBN 978-3-030-96592-1   (eBook)
https://doi.org/10.1007/978-3-030-96592-1

This Springer imprint is published by the registered company Springer Nature Switzerland AG
The registered company address is: Gewerbestrasse 11, 6330 Cham, Switzerland

Preface

Welcome to the proceedings of the International Workshop on Artificial Intelligence for Knowledge Management, Energy, and Sustainability (AI4KMES 2021), which was organized jointly by IFIP WG 12.6 - Knowledge Management and IFIP WG 12.11 - AI for Energy and Sustainability. AI4KMES 2021 took place in conjunction with the 30th International Joint Conference on Artificial Intelligence (IJCAI 2021), which was held online due to the ongoing COVID-19 pandemic, as was the previous IJCAI conference in Yokohama, Japan. The virtual conference and social events were amazing, and even though they lacked human exchanges, energy, and serendipity, the events still facilitated discussion and the exchange of new ideas and research results.

The IJCAI 2021 program was very rich including tutorials, 38 workshops, and panels. However, there were only a few communications on knowledge management and on sustainability, touching on components such as knowledge representation, dynamics of knowledge, knowledge base, knowledge transfer, shared knowledge, knowledge engineering, visual knowledge, and combining knowledge with deep convolutional neural networks. Proceedings from the main conference are available at https://www.ijcai.org/proceedings/2021/.

Knowledge management (KM) is a large multidisciplinary field having its roots in management and artificial intelligence (AI). Knowledge is one of the intangible capitals that influence the performance of organizations and their capacity to innovate. Since the beginning of the KM movement in the early 1990s, companies and nonprofit organizations have experimented with various approaches. AI has brought new ways of thinking, knowledge modeling, knowledge processing, and problem-solving techniques.

Understanding the benefits of knowledge management for research, organizations, and businesses and applying it is still a challenge for many. This is the same for sustainability, which should be considered from various perspectives. The overall process involving people, big data, and all kinds of computers and applications has the potential to accelerate discovery and innovation by organizing and optimizing the flow of knowledge. This collection of selected extended and revised papers from AI4KMES 2021 aims to challenge researchers and practitioners to better explore all AI fields and integrate feedback from real-world experience.

The first International Workshop on Artificial Intelligence for Knowledge Management (AI4KM) was organized by the International Federation for Information Processing (IFIP) Working Group 12.6 (Knowledge Management) in partnership with the European Conference on Artificial Intelligence (ECAI) in 2012, and the second workshop was held two years later during the Federated Conferences on Computer Science and Information Systems (FedCSIS) in conjunction with the 20th Conference on Knowledge Acquisition and Management (KAM 2014). The third edition of the workshop saw the beginning of the partnership with the International Joint Conference on Artificial Intelligence (IJCAI) in 2015. The fourth AI4KM workshop was held at

IJCAI 2016 in New York, USA; the fifth at IJCAI 2017 in Melbourne, Australia; the sixth at IJCAI-ECAI 2018 in Stockholm, Sweden; the seventh at IJCAI 2019 in Macao, China; and the eighth at IJCAI-PRICAI 20 in Yokohama, Japan.

With the aim of leveraging the new IFIP Working Group 12.11 - AI for Energy and Sustainability (AIES), this year's workshop was common to both AI4KM and AIES; hence, it was renamed the International Workshop on Artificial Intelligence for Knowledge Management, Energy, and Sustainability (AI4KMES).

The objective of this multidisciplinary conjunction is still to raise the interest of AI researchers and practitioners in knowledge management and sustainability challenges, to discuss methodological, technical, and organizational aspects of AI used for solving complex problems in these areas, and to share feedback on addressing these challenges with AI.

The theme of AI4KMES 2021 was "facing environmental challenges", with the specific objective of using AI approaches and techniques to support and improve the management of sustainable energy systems within smart cities, smart facilities, smart buildings, smart transportation, and smart houses. The ultimate goal of the workshop was to make a contribution towards achieving some of 17 UN Sustainability Development Goals.

This volume contains a selection of extended and revised papers from the workshop. The selection process focused on new contributions in any research area concerning the use of all AI fields for knowledge management, energy and sustainability. An extended Program Committee then evaluated the final versions of the papers, leading to this volume. We would like to thank the members of the Program Committee, who reviewed the papers and helped put together an interesting program. We would also like to thank all authors and our invited speakers, Rosiane de Freitas and Mieczyslaw Lech Owoc. Our thanks also go to the local Organizing Committee and all the supporting institutions and organizations. A summary of the papers included in this volume is provided below.

Our first invited talk by Rosiane de Freitas addressed a timely topic, "Applying AI/OR techniques and HR remote sensing to estimate how much the Amazon rainforest is helping us mitigate our carbon footprint". She presented joint research work with the National Institute for Amazonian Research, through the Forest Management Laboratory (LMF-INPA), and the Optimization, Algorithms, and Computational Complexity research group of the Institute of Computing, Federal University of Amazonas (ALGOX-IComp/UFAM). This project is partially supported by the Brazilian development agencies CNPq, CAPES, FAPEAM and led by Rosiane de Freitas (ALGOX-IComp/UFAM), João Marcos Bastos Cavalcanti (IComp/UFAM), and Niro Higuchi (LMF-INPA).

Based on the directives given by the Intergovernmental Panel on Climate Change (IPPC), there is an urgent need to provide additional guidance on the design of forest monitoring systems including issues such as forest inventory design, stratification, sampling, pools, accuracy/uncertainty assessment, and the combination of ground-based inventories with remote sensing and modelling approaches. The project concerns estimating carbon stocks by means of extrapolation and spatialization based on ground-based forest inventory combined with heterogeneous sources of remote sensing images through high-resolution satellite, radar, and LiDAR (Light Detection and Ranging) 3D technology. One of the objectives consists of determining a set of representative trees — the widest (dominant) and the highest (emerging) — through the application of computational intelligence strategies involving pattern recognition, data mining, graph theory, image retrieval, machine learning, and combinatorial optimization techniques. To refine the carbon stock estimate, complementary research consists of the detection of clearings in the Amazon forest based on satellite and radar images using machine learning techniques, which is a valuable tool in environmental conservation. However, small-scale clearing is a challenge. This is a recent trend hindering detection by satellite monitoring.

The first and opening presentation "Greening and Smarting IT – Case of Digital Transformation" gave an overview of multiple facets of the topics discussed with a common dominator: technology. Planet protection and climate change actions often neglect the pollution induced by technology and innovation. The Green Deal and Deep-Tech Digital Transformation (DT) are among the strategic topics of the European Union program Horizon Europe. Digital transformation offers a great opportunity for disruptive innovation- the consideration of the environmental aspects in the whole process may help to better control the impacts. Most eco-innovation actions focus on smart transportation, smart use of energy and water, and waste recycling but do not consider the necessary evolution of behaviors. While technology produces a twofold effect – benefits and waste, the capacity of available technology and in particular artificial intelligence for related complex problem solving is underexplored. The objective of this research is to provide an AI-based methodology for Smart and Green Digital Transformation. This paper explains how the knowledge-based AI and connectionist approaches and techniques combined with adequate thinking may innovate the management of DT in industry, administration, and other contexts and improve the effectiveness of efforts in sustaining the planet.

The paper entitled "Crowdsourcing and Sharing Economic in the Smart City Concept. Influence of the Idea on Development and Urban Resources" considers the impact of the sharing economy on smart cities.

The sharing economy is a relatively new trend that involves a complete change of organizational and distribution models. The main structure focuses on a distributed network of people and communities. This includes mutual service provision and sharing. Crowdsourcing is a process related to the sharing economy in terms of obtaining information/knowledge. The paper analyzes the relationship between these two trends and their impact on smart cities. Verification is carried out using the example of the city-state of Singapore, which has occupied leading places in the smart city rankings in recent years. An interesting aspect discussed in the paper is also the exchange of knowledge in a smart city and the cycle of information between residents, decision-makers, and stakeholders and third parties, such as education providers or businesses.

The next paper, "Assessment of Smart Waste Management with Spherical AHP Method", addresses the complex problem of waste management in a smart city context. Lack of a clear criterion in the evaluation of a smart waste management system (SWMS) required the use of multi-criteria decision making (MCDM) methods, allowing evaluation of both objective and subjective criteria. As evaluation needs both tangible and intangible data, using fuzzy logic is a useful tool to solve the problem.

Therefore, the spherical fuzzy analytic hierarchy process (AHP) method is used to handle the determined problem. Three alternatives are evaluated under four determined criteria. Results show that the problem is handled with spherical fuzzy AHP method by efficiently and effectively.

The authors of the next paper deal with "Zero Carbon Energy Transition in the Kitchens". They propose to combine carbon emission reduction and energy efficiency by using a carbon tax system within the jurisdiction of local authorities to transform cooktop ovens in kitchens in the South Atlantic region, selected for high propane usage. The carbon emissions are reduced by 1.2% with the proposed optimization model using the Residential Consumption Survey (RECS) data set. Based on the benefits of the first application, a new model is developed to look at the future with increasing demand. A regression-based machine learning is used in the R software to create a general model that predicts the efficiency increases. The model is constructed to assume that 100% of the propane cooktop ovens are converted into electric induction cooktop ovens. The authors expect that the proposed model will encourage the replacement of propane cooking devices with energy-efficient electric induction cooktop ovens to reduce carbon emissions and energy accessibility will be increased as energy-efficient appliances will be donated to the users by using the budget created through the carbon tax incomes.

The paper "Barriers and Challenges of Knowledge Management in a Gas Company" points out the potential barriers to the development of knowledge management in a gas company. Gas is a greener alternative to coal and it is used to heat houses as well as fuel cars. The growing gas industry market is dominated by large companies, operating across countries and continents. Managing such companies is extremely complex due to the scale of operations, rapid development of technologies, and a large number of employees working remotely and geographically dispersed. In the context of the increasingly common treatment of knowledge as one of an organization's key resources and the growing amount of data generated by business, knowledge management has received considerable attention in the energy sector, including the gas sector, because of its impact on performance. However, organizing the whole KM system is not an easy task because of complexity of whole process.

The next paper "Characterization of Residential Electricity Customers via Deep Ensemble Learning" explains the household characteristics in an electric grid and the importance of this data for the electric retailers, allowing them to provide personalized services, improve the demand response, and develop better energy efficiency programs. To avoid gathering the privacy-sensitive data, the authors propose an alternative solution exploring the electricity consumption data to infer household characteristics using supervised learning methods. The features are extracted from the electricity consumption patterns, and the selected features are used to train a classifier or regressor. However, the existing methods depend on a single contributing model, which can possibly be undertrained. To achieve the optimal performance of classifiers for characteristics identification, the authors propose an ensemble framework based on bagging algorithms.

"Grid Imbalance Prediction Using Particle Swarm Optimization and Neural Networks" deals with fluctuations in power demand. The imbalance costs are reflected in consumer prices in the partly liberated markets of developing countries. Thus, the accurate short-run forecast of electricity market trends is beneficial for both suppliers and utility companies to balance the physical energy supply and commercial revenue. When both the day-ahead market and the intra-day market exist to respond to the power demand, forecasting the imbalances assists both the suppliers and the regulators. This study aims to optimize the grid imbalance volume prediction by integrating particle swarm optimization (PSO) and long short-term memory recurrent neural networks (LSTMs). The model is applied for forecasts 1 hour, 4 hours, 8 hours, 12 hours, and 24 hours ahead. The mean absolute percentage error (MAPE) is also calculated. As a result, the MAPE levels are found to be 27.41 for 24 hours, 25.66 for 12 hours, 26.77 for 8 hours, 25.39 for 4 hours, and 9.25 for 1 hour. Although improvements are foreseen both in the model and in the data, the outcomes of this study would reduce the imbalance penalties for power generators and enable regulators to manage outages with a more precise approach. Hence, the economic benefits will affect the trading prices in the long term.

As our wish was also to introduce biomimetics, our second invited speaker and bee keeper, Mieczyslaw Lech Owoc, delivered an inspiring talk on the "Collective Intelligence of Honey Bees for Energy and Sustainability".

From the very beginning, the most promising AI methods have been inspired by the human environment and nature. In particular, the collective intelligence of non-human societies can surprise researchers and developers of new solutions. It is a matter of specific abilities of particular species oriented on cooperation and, moreover, awareness of precisely defined goals and resources used in very optimal ways. For example, we may admire honey bees' methods of building cells and organizing their work, where the problems of task planning and optimization of pollen collection are being solved through a certain sort of common intuition and collective intelligence. The proposed bees algorithm (as an example of swarm algorithms) has been applied in continuous domains (optimization of neural networks) or combinatorial ones (scheduling jobs for a machine). The goal of this paper is to present applications of collective intelligence in relatively new directions: energy acquisition and selected processes assuring sustainable development. Both directions seem to be very innovative and promising – especially in the ecosystems context.

The next two papers are devoted to nuclear plant safety. The first is entitled "The Application of Artificial Intelligence to Nuclear Power Plant Safety".

Through application of artificial intelligence (AI), the burden of analytical computational load for analysis of any given problem where countless variables have to be taken into account is virtually eliminated. Since for engagements in real life operations and instantaneous actions are of paramount importance, AI can be a strong alternative to overcoming complex problem solving in short time frames. As such, in this study a brief review of AI basics is given and literature for AI applications in the nuclear field such as defect detection in the nuclear fuel assembly, dose prediction in nuclear emergencies, fuel and component failure detection, core monitoring for reactor transients, core fuel optimization models, gamma spectroscopy analysis, and, specifically, nuclear reactor safety in operation are assessed. Afterwards, an AI model for analyzing transients in the VVER type nuclear power plants that are being built in Turkey is proposed. This model must keep up with instantaneous data flow and give actionable feedback to the operator for both the cause and the solution of any problem.

A semi-autonomous AI control system that helps operator decision making is a significant contributor to the safety of a nuclear reactor.

The authors of "Capacity to Build Artificial Intelligence Systems for Nuclear Energy Security and Sustainability: Experience of Belarus" address a new type of innovation, called "situational innovation", essential for ensuring people's health and safety. As in recent decades the danger of man-made hazards in the nuclear field has increased dramatically, Belarusian scientists have accumulated considerable innovative potential in health care, agriculture, and the creation of new life support technologies in radioactively contaminated areas. This information is required when working out catastrophe scenarios at nuclear power plants. The definition of situational innovations is developed. Highlighting situational innovations makes it possible to determine their adaptation, use, and replication in other economies, to diffuse the knowledge, and to increase cost effectiveness. The results of the study indicate that the unique experience of Belarus in overcoming the consequences of a nuclear disaster is of great practical importance for other countries operating nuclear objects. The situational innovations resulting from such disasters need to be consolidated into an international database of nuclear research results. The application of artificial intelligence (AI) systems allows the accumulation, storage, and retrieval of information in a single source and would provide the necessary situational innovations in the case of a nuclear accident. AI systems also help to prevent and reduce the risk of nuclear accidents. Development and use of AI will allow countries worldwide to develop nuclear disaster information management systems and reduce existing disaster risk for sustainable development in the future.

Smart electrical grids also play a major role in energy transition but raise important software problems. The authors of "Automated Planning to Evolve Smart Grids with Renewable Energies" propose to solve some of them using AI techniques. In particular, the increasing use of distributed generation based on renewable energies (wind, photovoltaic, among others) leads to the issue of integration into distribution networks that were not originally designed to accommodate generation units but to carry electricity from the distribution network to medium and low voltage consumers. Some methods have been used to automatically build target architectures, to be reached within a given time horizon (of several decades), capable of accommodating a massive insertion of distributed generation while guaranteeing some technical constraints. However, these target networks may be quite different from the existing ones and therefore a direct mutation of the network would be too costly. It is therefore necessary to define the succession of works year after year to reach the target. The authors address this by translating it to an automated planning problem. They define a transformation of the distribution network knowledge into a PDDL representation. The modeled domain representation is fed to a planner to obtain the set of lines to be built and deconstructed until the target is reached. Experimental analysis, on several networks at different scales, demonstrates the applicability of the approach and the reduction in reliance on expert knowledge. The objective of further work is to mutate an initial network towards a target network while minimizing the total cost and respecting technical constraints.

The paper "Artificial Intelligence Application for Crude Distillation Unit: An Overview" focuses on crude distillations units (CDU), for which energy optimization has been a tremendous challenge because of their complexity. The presented overview shows that soft sensors are the most common application of artificial intelligence for a CDU, although a number of recent publications focus on optimization problems. The approaches for optimization are very diverse, which makes them difficult to apply in the current engineering practice. This work provides a guideline for selecting the right method, but also addresses the fact that different methods excel at different problems and with different data set sizes. For neural networks (NN), this further depends on their architecture and hyperparameter adjustment. This urges future research, where the goal could be a workflow that would automatically adapt methods and perform parameter tuning with minimum user input.

In "Deep Reinforced Learning for the Governance of a Sample Microgrid" the authors propose a proximal policy optimization reinforcement learning system to handle the energy dispatch management of a sample microgrid. The considered microgrid has three participants of different classifications, signifying their relative importance and how sensitive they are to energy shortages. The energy within the microgrid is generated by these participants, who are individually equipped with a solar panel and a wind turbine for energy generation, along with an energy storage system. The environmental conditions, i.e. temperature, wind velocity, and irradiation figures for Istanbul, are considered to obtain accurate energy generation figures. The microgrid is designed to be grid connected in order to compensate for the uncertainties caused by the weather changes, and hence utility service is accessed when the energy produced and stored cannot respond to the demand. Information security of the participants is respected and to that end, direct energy generation, consumption, and storage figures are not supplied to the agent, instead only supply and demand figures are transferred. The agent, using this information, after a period of training, optimizes the system for a reward scheme that rewards energy exports and punishes energy deficits and imports. The results verify the feasibility of proximal policy optimization in managing microgrid energy dispatch.

The authors of the paper "Residential Short-Term Load Forecasting via Meta Learning and Domain Augmentation" observe that with the increasing adoption of electric devices and renewable energy generation, electric load forecasting, especially short-term load forecasting (STLF), has recently attracted more attention. Accurate short-term load forecasting is of significant importance for the safe and efficient operation of power grids. Deep learning-based models have achieved impressive success on several applications, including short-term load forecasting. Yet, most deep learning models require a large amount of training data. However, in the real world, it may be very difficult or even impossible to collect enough data to train a reliable machine learning model. This makes it hard to adopt deep learning models for several real-world scenarios. Thus, it will be very helpful if deep learning models can be learned to tackle tasks with a limited amount of training data and unseen tasks. The authors propose to use the meta-learning framework to train a long short-term memory-based model for short-term residential load forecasting. Specially, by minimizing the task-level loss (loss over several tasks), the model is trained to perform well on different tasks. They also use domain randomization techniques to further augment the training tasks, which may further improve the generalization ability of the proposed model. The proposed model is evaluated on real-world data sets and compared against some classic forecasting models.

The increasing urban population creates escalating problems such as housing, infrastructure, transportation, health, environment, safety, and energy consumption. Climate change, emission mitigation, and limited energy supply force urban managers to consider sound measures with the support of technological developments. This is based on data collection and accumulation using IoT, sensors, digital networks, and other means. "Smart urbanism" is a concept that considers predicting, designing, and creating solutions in a systematic, sustainable, and agile manner based on the data collected. Energy is an indispensable dimension in this context. Optimum energy management makes it "smart energy" with the inclusion of clean and sustainable renewable energy resources as well as energy efficiency. In "Renewable Energy Investment Decision Evaluation for Local Authorities" the possible inclusion of geothermal, solar, and wind power is analyzed to identify the best feasible alternative considering the parameters of location, climate, space availability, and capital and operational expenditures as well as construction, operation, and maintenance. The fuzzy analytic hierarchy process (AHP) technique is used to evaluate the ranking. The proposed method uses fuzzy mathematics for solving problems containing uncertainties as well as less quantifiable inputs. This study proposes a methodological framework for the analysis of competitiveness of alternative renewable energy generation in urban environments. The municipality of Balıkesir is chosen for the case study presented in this work.

We hope you enjoy reading these papers.

January 2022   Eunika Mercier-Laurent
                                                                        Gülgün Kayakutlu

Organization

Program Chairs

Eunika Mercier-Laurent   University of Reims Champagne-Ardenne, France
Gülgün Kayakutlu   Istanbul Technical University, Turkey

Program Committee

Danielle Boulanger   University of Lyon 3, France
Burak Barutcu   Istanbul Technical University, Turkey
Anne Dourgnon   EDF Research Center, France
Otthein Hertzog   Jacobs University, Germany
Knut Hinkelmann   University of Applied Sciences and Arts, Switzerland
Gülgün Kayakutlu   Istanbul Technical University, Turkey
M. Ozgur Kayalica   Istanbul Technical University, Turkey
Antoni Ligeza   AGH University of Science and Technology, Poland
Vítězslav Máša   Brno University of Technology, Czech Republic
Nada Matta   Troyes Technical University, France
Eunika Mercier-Laurent   University of Reims Champagne-Ardenne, France
Mieczyslaw Lech Owoc   Wroclaw University of Economics, Poland
Maciej Pondel   Wroclaw University of Economics, Poland
Vincent Ribiere   IKI, Thailand
Michael Stankosky   George Washington University, USA
Abdul Sattar   Griffith University, Australia
Frederique Segond   Inria, France
Senem Şentürk Lüle   Istanbul Technical University, Turkey
Guillermo Simari   Universidad Nacional del Sur, Argentina
Hiroshi Takeda   Osaka University, Japan
Mario Tokoro   Sony-CSL, Japan
Eric Tsui   Hong Kong Polytechnic University, Hong Kong
Janusz Wojtusiak   George Mason University, Fairfax, USA
Berker Yurtseven   Istanbul Technical University, Turkey

Local Organizing Committee

Min-Ling Zhang   Southeast University, China

Contents

Greening and Smarting IT – Case of Digital Transformation

Eunika Mercier-Laurent[✉]

University of Reims Champagne Ardenne, Reims, France
eunika.mercier-laurent@univ-reims.fr

Abstract. The recent concern about the Planet condition and climate change does not include in the debates the pollution induced by technology and innovation. Green Deal and Deep-Tech Digital Transformation (DT) are among the strategic topics of the European Union program Horizon Europe. They cover various fields that are not fully represented in the calls for proposals. Digital transformation offers a great opportunity for disruptive innovation, but consideration of the environmental aspects in the whole process may help better controlling the environmental impact. Eco-innovation actions focus mainly on smart transportation, smart use of energy and water and waste recycling but do not consider the necessary evolution of behaviors. While technology produces twofold effect – benefits and waste, the capacity of available technology and in particular Artificial Intelligence for related complex problems solving is underexplored.

The objective of this research is to provide the AI-based methodology for Smart and Green Digital Transformation. This paper explains how the knowledge-based AI and connectionist approaches and techniques combined with adequate thinking may innovate the way of managing DT in industry, administration and in other contexts and improve the effectiveness of efforts in sustaining the Planet.

Keywords: Artificial Intelligence · Digital Transformation · Method · Sustainability · Sustainable development · Planet crisis · Climate change · Data obesity

1 Introduction

The participants of recent COP26 aim in accelerating action towards the goals of the Paris Agreement and the UN Framework Convention on Climate Change. Increasing frequency and strength of natural disasters is a complex consequence of decisions taken at governmental and other levels and of human behaviors as well.

From the beginning of intensive industrialization after the second war some alerts have been raised by among others Antonina Lenkowa [1] and Erik Eckholm [2] trying to explain to governments and people the various and heavy consequences of such intensive industrialization.

Since the world population has tripled. Globalization amplified by technological progress leads to increased transportation of goods and of persons. Many focus on

© IFIP International Federation for Information Processing 2022
Published by Springer Nature Switzerland AG 2022
E. Mercier-Laurent and G. Kayakutlu (Eds.): AI4KMES 2021, IFIP AICT 637, pp. 1–18, 2022.
https://doi.org/10.1007/978-3-030-96592-1_1

business only. Despite the large introduction of Corporate Social Responsibility, some companies still practice planned obsolescence to generate more revenues [3].

Is getting huge and quick revenue compatible with Planet protection?

Technological progress facilitates communication, sales online, but also generates lot of visible and hidden waste [4]. Numerous reports elaborated by WHO [5], OECD [6] and GIEC [7] and conferences including the recent COP26 alert again about the state of our planet Earth. Intergovernmental Panel on Climate Change (IPCC), the UN body was created to provide policymakers with regular scientific assessments on climate change, its implications and potential future risks, as well as to put forward adaptation and mitigation option. It is composed of several working groups including one multidisciplinary. Through its assessments, the IPCC determines the state of knowledge on climate change. It identifies where there is agreement in the scientific community on topics related to climate change, and where further research is needed [8].

However, the involved stakeholders focus mainly on carbon dioxide generation, but it is only one factor. Main conferences devoted to climate change blame coal and automotive industry as the biggest emitters of carbon dioxide. They seem ignore the other influencing factors such as globalization and miss to study the pollution systems as a whole.

In reality the addressed problem is very complex and effective solving it requires collaboration of specialists from various fields, deep understanding and actions, not only recommendation and reports.

While the information technologies including artificial intelligence are ubiquitous they are weakly considered as a source of pollution. Computers and other devices are designed to be changed often. Data centers generate heat due to increasing data obesity. There is a lack of "garbage collector" cleaning outdated, duplicated and false data.

Advertisement-based business model produces "invisible" pollution via cookies, not targeted emails, social networks influencers' and other trends.

On the European level the program Horizon Europe addresses Digital Transformation and Green deal separately; submitted projects are evaluated by experts in related fields what limits the opportunity for disruptive innovation. The recent posts on twitter show emerging synergy between two programs [9].

After discussion on the main contribution of IT to Planet Crisis, the case of Digital Transformation and Green Deal are analyzed applying ecosystem approach. Some elements of AI-based methodology are proposed and commented before the final perspective for the future work.

2   Sustainability and Climate Change – Current Efforts and Influencing Factors

In general manner sustainability is about the integration of environmental health, social equity and economic vitality in order to create thriving, healthy, diverse and resilient communities for this generation and generations to come. The practice of sustainability recognizes that these complex issues are interconnected and requires a systems approach. Indeed sustainability depends on how human activities are in balance with natural ecosystems and how they affect biodiversity and our biosphere [4].

The International Standard ISO 26000 provides guidelines on how businesses and organizations can operate in a socially responsible way [10]. While it constraints companies to integrate social and environmental aspects into their activities, for ex. design, they should also focus on balancing it with their economic performance. These guidelines are complex and it is very difficult for a small company to check and respect all these principles without losing business.

Introducing the environmental principles into a design of electronic devices and related services is a step forward. However, the traditional PLM (Product Life Management) tools should evolve to take into consideration a new way of doing [11] by using alternative thinking, simulators before doing and optimizers. Future Factories are supposed to evolve in this direction [12, 13].

2.1 Influencing Factors

Globalization, innovation and IT are weakly considered as strong influencing factors by all conferences and debate on Climate Change.

2.1.1 Globalization

Focus on profit "World companies" have changed the economic landscape. Opening to business and quick development of China and other Asian and South American countries offering the low labor cost increases the relocation-out from origin countries, mainly US and Europe, in search of quick and high revenues. Such business relies on not always optimized transportation across the world by ships and airplanes, which increases pollution [14]. Besides, the end-buyers have to recycle products, often of poor quality, made just for sale. Repairing them costs often more than new product and requires the availability of spare pieces and related tools and expertise. Many integrate embedded planned obsolescence.

Huge global business amplified through communication technology, e-commerce, and transportation facilities. The each day of recent Suez Canal blockage disrupted more than $9 billion worth of goods, according to The Associated Press, citing estimates from Lloyd's List. The large number of ships stuck in one place—at the northern end of the Suez Canal in the Mediterranean Sea—caused Sulphur emissions to spike in the region [15]. Route through North Pole will increase of ice melting.

The knowledge of adequate techniques of AI and expertise in applying them can help finding and optimizing the local resources in aim avoiding or minimizing transportation of many goods and enhance local economy. AI is also useful for optimizing logistics. Covid pandemics demonstrated clearly the need for relocation in, especially for electronic components in Europe.

2.1.2 Innovation

Despite the provided benefits, the innovation is among contributors to planet disaster, because the inventors and designers think mostly about the functionalities, shape, look, attractiveness, and few about minimizing the use of raw materials, reducing the energy consumption, providing the remanufacturing and recycling facilities and other ways to reduce environmental impacts. "Innovation" in packaging by reducing the quantity of a given product (food, ink) in more attractive box produces more waste.

The primary eco-innovation movement claims to generate new businesses and jobs but the use of technology in the related activities is moderate, except maybe for waste sorting [16]. Instead of sorting, it is smarter to produce less waste. The standards associated with product lifecycle management (PLM) software claim to manage environmental impact of the design [17]. Autodesk is probably among the pioneers of industrial convergence including constraints into design [18]. However, they do not consider yet the environmental and recycling constraints.

The industrial renewal in Europe called Industry 4.0, initiated by German government to face Asian competition, focuses on digitalization and optimization of related processes. Therefore, eco-design is included into Industry 4.0 [19]. Future factories, such as those at Vaudreuil of Schneider Electric and Safran in France integrate a lot of AI, preserving principle of collaboration human-machine, instead of replacing humans. Cyber physical systems allow simulate and improve maintenance of existing equipment and design [12, 13].

Technological progress, considered as powerful engine of economy produces twofold effect. It provides benefits for humanity in many fields, but contributes also to Planet Crisis with massive waste generation, energy consumption, generation of heat by data centers situated mainly in the north and depletion of scarce metals. The enthusiasm of the users for novelty and recently for AI lead to sometimes hidden waste.

3 Twofold Effect of Information Technologies

Information technologies have provided a considerable help for humans in various contexts, such as administration, businesses, schools, research, health, cities and many others. Everything digital and connected is everywhere, even if it is not necessary. Thanks to IT many activities went online during the Covid pandemics. This experience demonstrated that it is possible to reduce traveling and optimize between telework and work in office. Although records from meeting and conferences are registered and stored contributing to data obesity.

Related to innovation, technology is able to enhance creativity (generator of ideas in [20]) and provide a considerable help in eco-design, verifying constraints before implementation, optimizing design, reparability and recyclability. However, the way of thinking compatible with such complex problem solving should combine the global, holistic and system approaches [4, 21]. For example, aerospace and automotive industries focus on lightening weight and reducing carbon footprint, while other related aspects such as route optimization for users are weakly considered. Current travel search engines works for their clients, not for travellers. The air transportation system "hub-and-spoke" in which local airports offer air transportation to a central airport where long-distance flights are available, need innovation to minimize the distance and related CO2 emission. The effort can be done through better collaboration between airlines. Three main alliances Sky Team, Star Alliance and One world do not accept combining the flight tickets between them for optimizing flights distance.

The AI-based simulators can help checking and optimizing things. Such system has been prototyped in the framework on French national funded project Convergence [17] that delivered a game-based guide for SMEs aiming in guiding them in applying ISO 26000 requirements related to eco-design.

Amazon claims reducing carbon dioxide emission by using technology in the whole cycle of sales [22]. Although, they also contribute to C02 generation. Total balance is not published.

The initiatives as buy local, sharing knowledge on repairing via video and numerous applications, such as Uber, wonolo [23] and some others, connect people offering services or jobs with those who need them. It is quick and effective. Some platforms connecting offer and demand use AI techniques such as Case-based reasoning or deep learning for direct matching (e-commerce, job finding, talent finding, travel, help desk) [24, 25].

Games and serious games (with AI) have also potential to encourage people to think and act differently. For example, the game Nega-water developed by association Du Flocon à la Vague calculates the use of water when doing daily activities [26].

However, the majority of computers, smartphones, IoT and other electronic devices are mostly not eco-designed, except Xiaomi Modular smart phone (see Fig. 2) and even integrate a principle of planned obsolescence. Consequently they are thrown away and replaced by the latest models. Not conceived with knowledge approach, the consecutive layers of software and frequent update for various reasons increase already existing "obesity" and generate a need for more powerful computers, smartphones and other devices. The combination of various communicating software requires "up-to-date" hardware to run correctly. Software is so complex that it is impossible to remove not-used parts (ex.Windows).

While web 2.0 brought useful services, the massive appropriation of social networks and various applications produce an exponential amount of data stored in data centers that need cooling. Fortunately, few apply circular energy to reduce environmental impact [30], but still those in Scandinavian countries and North America clearly contribute to the rise of temperature and melting of ice [4, 29].

The third hype of AI has been mostly triggered by marketing needs to sale more and quicker to largest public. The involved companies collect all available data in aim to elaborate models and selected algorithms to support the whole acceleration and amplification of business including "client experience". Covid 19 pandemics amplified also online shopping....and need for transportation.

The advertisement companies explore the navigation data and "client experience" in aim pushing not always targeted advertisement to large population of the Internet users. It generates not only carbon footprint, but also visual pollution and certainly influence concentration. According to [27, 28] simple email without attachment generates equivalent of 4 g CO2, annual google search 10 kg by only one person.

Advertisement-based business model, invented by Google influences and mass media promotes the "buy more" and "to have more and to show" mentality. "Buy more, throw away and buy new" are the engines of today business. The demand for data scientists, especially in marketing is increasing, which is good for employment.

The trend of "Data and machine learning" affected also the problem solving capacity among the AI programmers to the point that many apply Big Data + machine learning principle to solve all kind of problems even if the alternative methods, greener and smarter are available. Collection of data became a prerequisite. However, their quality should be adapted to the expected results without cheating aiming in influencing the results and producing the famous bias. Raw data need "cleaning" before processing.

In fact, the quality of data is not controlled – there is not "garbage collector" for outdated data, no verification of consistency, and no selection in function of target. For example, many organizations collect our personal data that are in various database instead of having just one (identity). By consequence the same data may be registered a dozen of times, contributing again to data obesity.

Artificial Intelligence plays already a significant role in decision support systems, optimization, simulation, but can do better at the condition to consider the mentioned elements as an ecosystem. It requires a capacity of different thinking without the barriers between fields [31], complex problem solving, art of choosing appropriate techniques and managing available knowledge and experience that apply.

Digital Transformation and Green Deal European programs provide an opportunity to do things differently.

4 Digital Transformation and Green Deal

European Union strategy includes two ambitious programs: Digital transformation [32] and Green Deal [33]. The first includes Industry 4.0, next generation of aircraft, transformation in design, manufacturing, integration and maintenance. Green Deal focus mainly on energy, water, transportation, among others.

However these two topics are interrelated.

4.1 Digital Transformation

According to IT businesses Digital transformation is the process of using digital technologies to create new—or modify existing—**business processes, culture, and customer experiences** to meet changing business and market requirements. This definition clearly focus on business and lacks of considering environmental aspects.

Oracle mentions five large fields concerned by digital transformation: manufacturing, retail, healthcare industry and smart cities.

"In manufacturing, IoT can improve efficiency and profitability by enabling predictive maintenance and process improvements that reduce downtime and scrap while ensuring the production of high-quality goods. Augmented reality solutions enable superior and less-costly training and can bring expertise from afar directly into the factory.

In the retail environment, digitalization leverages devices, channels, and platforms to create seamless customer experiences tailored to specific customer needs. Chatbots, AI, and sophisticated data analytics are helping retailers develop and deliver personalized recommendations to successfully appeal to an audience of one. Digitalization is transforming the in-store experience as well, creating an engaging interactive environment and more convenient customer service processes.

The healthcare industry is leveraging a variety of digital capabilities to create patient-centric and value-based outcomes that benefit patients, providers, and payers—while reducing costs at the same time. Virtual doctor visits and networked electronic medical records are just two modes of digitalization that are transforming healthcare.

Digital-enabled "smart cities" that blend existing physical infrastructures with cutting-edge digital technologies are becoming more efficient while providing a safer and better quality of life for their citizens. From utility monitoring to public safety to environmental sustainability, technologies such as sensors, AI, and video analytics are helping the public sector transform the way it provides essential services to achieve greater efficiency, lower costs, and a higher level of citizen engagement" [34].

Such an approach leads to more IT waste and certainly to energy waste. Everything is based on data business.

MIT limits Digital transformation to block chain, IOT, cloud and cybersecurity [35 MIT].

European Union elaborated Digital Europe Program designed to bridge the gap between digital technology research and market deployment. It aims in bringing benefit to Europe's citizens and businesses, especially SMEs. Investment under the Digital Europe Program supports the European Union's twin objectives of a green transition and digital transformation while strengthening the Union's resilience and digital sovereignty [36].

The special focus is on Industry 4.0 (digital twins), next generation aircraft, digital transformation in design, manufacturing, integration and maintenance.

Public administration seems to be forgotten, while there are a lot to do including the way of thinking. Usually people has adapt to existing software, "fat" for the most. The smart and green transformation requires evolving the traditional way of thinking and designing application in collaboration with the users [20, 37, 38, 39].

The current call for proposals The Digital Europe Program [40] provides funding for projects in five crucial areas:

- supercomputing
- artificial intelligence
- cybersecurity
- advanced digital skills
- ensuring the wide use of digital technologies across the economy and society

Supercomputing is necessary because of massive generation of multimedia data. Artificial Intelligence is not a topic per se, but AI approaches and techniques should be integrated in all above points. Cybersecurity requires also AI thinking in terms of understanding hackers methods and maybe modifying their bad attitudes for ex. via games.

Related to skills, the first thing to do is to know what we have and what we need and plan training and education accordingly [41].

While health systems are absent in this call for proposals "digitalization and the use of digital technologies is already an important and growing part of most health systems. However, national governments and international health partners are not yet fully unlocking the potential of digital health to scale up access to primary health care services as a pathway to achieve UHC" [42]. The strategy for prevention is also missing.

All mentioned requires *"smart & green IT" thinking* [43]

4.2 Green Deal

Another EU program: Green Deal is represented in Fig. 1.

It covers nine areas: climate ambition, clean and secure energy, industry and circular economy, efficient buildings, mobility, agriculture, biodiversity and very ambitious zero pollution. Three horizontal areas are: strengthening of knowledge to support the 9 areas, empowering citizens in transition to climate neutral sustainable Europe and international cooperation with Africa and Mediterranean.

Fig. 1. Source [33]

The program ambitions in "transforming the EU into a modern, resource-efficient and competitive economy, ensuring: no net emissions of greenhouse gases by 2050, economic growth decoupled from resource use, no person and no place left behind".

Expected benefits:
fresh air, clean water, healthy soil and biodiversity
renovated, energy efficient buildings
healthy and affordable food
more public transport
cleaner energy and cutting-edge clean technological innovation
longer lasting products that can be repaired, recycled and re-used
future-proof jobs and skills training for the transition
globally competitive and resilient industry

These are just wishful thinking. The actions mentioned on GD website are all connected. Technology is supporting all these areas, but technology is not represented as area [44, 62].

Effectiveness of circular economy is related to eco-design, packaging and depends on business model. Reusing the components requires adequate design. For ex Xiaomi elaborated a "lego-like" smartphone, shown in Fig. 2.

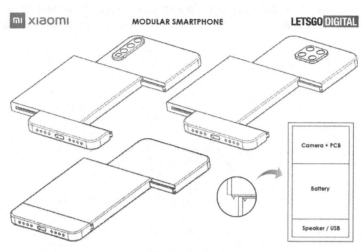

Fig. 2. Source.

European eco-design preparatory study on mobile phones, smartphones and tablets proposes a methodology for their eco-design [45].

Extending the principle of circular energy to data centers will be certainly beneficial for Planet and energy users.

Clean energy and water, efficient building houses, public transportation are already considered in Smart City programs and offers [43].

The smart use of AI approaches and techniques may help achieving some of above mentioned objectives.

5 Contribution of Artificial Intelligence to Smart and Green Digital Transformation

Since its beginning AI researchers and practitioners invented and experimented various AI approaches and techniques for solving complex problems [63, 66]. Most of these techniques shown in Fig. 1 in [43] have been successfully experimented for various applications in many fields around the globe. Decision support systems, diagnosis and many others are now embedded in larger applications such as automated pilot, autonomous car, intelligent building, in drones, translation systems, personal assistants, security and many others.

Knowledge-based AI introduced specific way of thinking and problem solving which is different from those related to exploring databases. For example, managing sustainability requires deep understanding of all influencing factors, relations between them and elements of context, such as policies and human behaviors. Applied to Digital Transformation such approach will be very helpful in smarting and greening IT.

The approach such as KADS [46] combing Newell's knowledge level, Chandrasekeran's interaction hypothesis and Clancey generalization, requires defining of a goal and decomposing it into the tasks; each tasks corresponds to a specific reasoning model that will use available knowledge and consider the context to achieve a given goal, see Fig. 3.

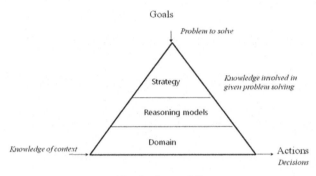

Fig. 3. Source [4]

Building a conceptual knowledge model consists in elaborating related ontology representing concepts and relations between them and reasoning models needed for given problem solving. Such models can be generalized and reused for other applications. It is similar to building a toy from Lego blocks. Modularity, genericity and reusability are 3 pillars of green and smart programming.

Although the third hype of AI is a comeback to data analytics that has been applied for years in industrial safety and reliability [47]. It consists in collecting data first and applying statistics or suited algorithms to "make data talk". Specialists in marketing were suddenly empowered by analytics and "machine learning" algorithms. As deep-learning requires big data for more accurate results they focus mainly on gathering more data from various available sources about customers to push them more products and services. Collected data is mined using analytics and deep learning software and the results explored commercially. This approach is not sustainable, because it requires a huge amount of data (data obesity) and consequently their storage and cooling, as well as energy for processing.

Third generation of AI includes also evolution of robots – now robots are deployed in many contexts without necessarily having a need for robot capacities. Currently many various robots replace humans including military robots and drones what raise ethical questions [48, 67]. Very few is wondering about sustainability, eco-designing of robots, their recyclability and reparability, availability of raw materials and of electronic components to build "super-robots".

An "automated" dialog between human and machine was proposed by Turing (famous Turing test) and Weizenbaum (Eliza 1966) [49], today there is a trend for having a chatbot able to answer the questions of clients and visitors. The designers of "intelligent" assistants implemented this principle. The majority of chatbots explore the ready answers recorded in (again) database, however the smarter chatbots, based on text mining or dialog understanding are coming [50].

Increasing computer power and on cloud storage facilities makes people recklessly producing large amounts of data. Only trendy techniques are used to solve every kind of problems, but the connectionist AI exploring data bases is not sufficient for smarting and greening IT. With main focus on prediction this approach is characterized by ignorance of past knowledge and experience for the quality of obtained results.

Since years, researchers and practitioners of AI have demonstrated the AI has potential to provide valuable solutions for facing today challenges. Since the 1980s, numerous applications in the fields of medicine, chemistry, industry, finance and others demonstrated that AI is able to help human in many situations [51].

The engineering specialists understood that in many cases it is necessary to combine "data approach" with "knowledge–based" AI for more accurate results.

5.1 What AI Can Do

Knowing the available AI "toolbox" and deep problem understanding combining KADS and global, holistic and system thinking facilitates modularity and reusability and the choice of right technique or combination of them in hybrid systems [20, 53]. For example, using constraint programming for optimizing, planning and scheduling is smarter and more effective than trying to solve it with other techniques. Artificial neural networks are effective to find similar pattern in images, text or in audio file, case-based reasoning for matching local offers of jobs and competency and in intelligent e-commerce application. Some of mentioned techniques are integrated in multi-agents systems (MAS) [52]. They have been used to study and to simulate complex systems in different application domains where physical factor are present for energy minimizing, where physical objects tend to reach the lowest energy consumption possible within the physically constrained world. Furthermore, MAS have been intensively exploited for the analysis through simulation of biological and chemical systems.

Achieving the ambitious goal of greening and smarting IT requires covering the whole cycle of live of devices and software. AI can support all steps of this cycle.

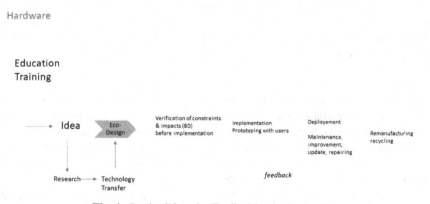

Fig. 4. Device lifecycle (Eunika Mercier-Laurent)

The Fig. 4 represents a lifecycle of green and smart hardware.

The process begin with an idea: new architecture, shape and miniaturization. In the case of research in general technology transfer of research results to industry is necessary. The next step is eco-design of architecture. The constraints such as market, feasibility, upgradeability, materials (easy recycling), environmental impact and some others have to be verified before implementation. Next steps are prototyping and deployment. It includes also maintenance and improvement in function of feedback provided. The eco-design has to include the end of life: remanufacturing or recycling.

A similar approach can be applied to AI software.

According to El Fallah Segrouchni [54], the AI lifecycle is composed of following steps: research, design, development, deployment, use, maintenance, conversion, dismantling and disabling. Applying of principle modularity, genericity and reusability makes the cycle smarter. Using the right tool from the AI Toolbox contributes to greening. For example, constraint programming is very useful for optimization, scheduling, resource or frequency allocation, timetable and logistics at the condition to know the constraints and the values they can take. It facilitates a real time reprogramming in case of new, strong constraint.

Education plays an important role in optimizing the whole process. Education and continuous learning are the basis for smarting and greening IT. Education should include very early the various ways of thinking and problem solving including finding alternative solution. As IT and AI are everywhere these approaches should be learned from small school. Followed latter by deep-understanding of complex problems and smart problem solving including considering alternative solutions [65].

AI-based simulation in VR environment effectively support quick prototyping. Simulation helps evaluation of impacts before implementation. An example of such simulation available on World Economic Forum website [55] is presented in Fig. 5.

This kind of simulation should be applied in many decision support systems. It will be especially useful for decision making at governmental level.

The main impacts of IT and AI are represented in Fig. 6. All are interrelated.

In the case of sustainability the contextual knowledge on policies in a given country, base of available actions and practice/prototyping should be considered. Taking inspiration from games [56] we can imagine building a Sustainability Support System as a simulator containing various interrelated components to evaluate the impact of decisions for policy makers, planners, organizations and companies. It can be conceived as a toolbox allowing easy configuration in function of the needs and allowing verifying various constraints before doing.

Reusing

Some of Smart cities opened their data to large public, what is a data recycling for various applications and data-based business. The EU Cordis offer the access to research results [57]; it can provide inspiration for digital transformation in various fields.

Re-using past knowledge and experience seems to be forgotten with ML approaches, thought it could help agricultures and achieving the UN SD goal nb 2 [58]. Technology is not included in 17 goals, but it can help achieving quite all of them [43].

According to Eric Thivant [59] all outdated electronic devices can be partially recycled or reused in the developing countries that lacks computers.

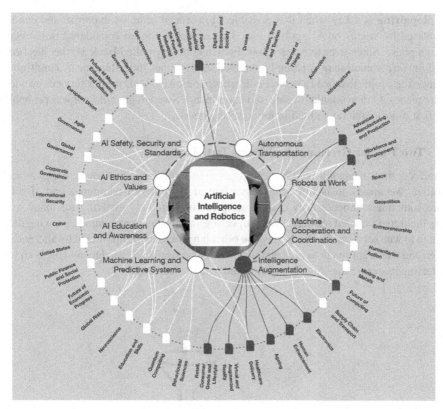

Fig. 5. Simulation of impacts, source https://intelligence.weforum.org/topics/a1Gb0000000p TDREA2?tab=publications

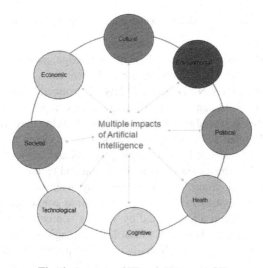

Fig. 6. Impacts of IT and AI, source [4]

Repairing is a key when it is possible. However, it requires expertise and can be supported by videos or AI applications. This aspect has to be considered in the eco-design phase. Since over five years now, French company called Back Market has been working on revaluation of apparently outdated smartphones [60]. Repair cafés multiplies connecting people that need a service or learn reparation [61]. 3D printer services provide spare pieces if not available for sale (old equipment). Sweden government was probably the first offering the tax reduction for repairing goods.

5.2 Two Tips for Greening and Sliming Data Base

AI affects the sustainability. Two aspects of AI should be investigated: negative and positive. Negative consists in being dependent of all forms of computers and of software sometimes unnecessarily complicated; the both quickly evolving for many reasons including planned obsolescence. Collecting and storing exponentially growing data related to web services, for marketing and other purposes can be optimized by alternative green and smart approach to programming, using the suited knowledge models, both for collecting relevant data (Fig. 7) and for guidance in "learning".

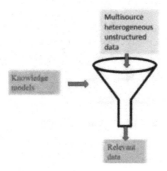

Fig. 7. Collection of data guided by knowledge models (Eunika Mercier-Laurent)

Another possibility is to combine DL with knowledge, as shown in Fig. 8, in aim providing explanations and offer knowledge-based services.

The achievement of the UN SD Goal 12 "Ensure sustainable consumption and production patterns" again AI can help preventing various impacts. Eco-innovation and eco-design approaches powered by AI and alternative thinking may lead to wise exploring of raw materials, even avoid using them, especially in electronic devices. Planned obsolescence has to be replaced by sustainable products and services.

Goal 17 "Strengthen the means of implementation and revitalize the global partnership for sustainable development" can be achieved by incentives such as investment in sustainable innovation applied to Digital transformation connected with Green Deal. Global partnership has been already tried by various conference but without significant results. Interesting initiatives such as Earth Planet, agro ecology of CIRAD [64], numerous projects on preserving biodiversity and other prototyping actions [may serve as inspiration for the next. Knowledge base BC3 [59] was designed to group such experiences.

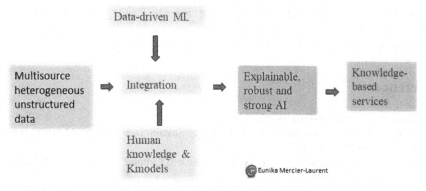

Fig. 8. Combining DL and K-based AI

Digital transformation can be a playground for prototyping of sustainable development of not only industry but also all fields concerned by Digital transformation.

Knowledge Management principle may evolving the way of thinking by asking the right questions: who are the contributors and users, how to organize whole process of data, information and knowledge collection, how to help connecting all elements and allow information and knowledge sharing, the observation of world best practices.

6 Conclusion and Perspective

Humans are the greater generators of disasters. In the context of Planet Emergency, sustainability and sustainable development is still a big challenge for humanity. Before applying technology this complex problem requires awareness about our impact and evolution of individual and companies behaviors, radical change of focus from having more to living better in harmony with nature.

To remain sustainable, the IT development requires a smart use of available AI techniques, not only trendy ones but also those effective for a given case. Combined with intensive use of available knowledge – individual and collective, from related domains, from the past and currently gained from experience such systems can really help minimizing impact of human activity.

AI way of thinking helps understanding better such a complex problem, finding alternative solutions. The better use of our brain capacity is mandatory.

AI is useful in many situations if we apply appropriate method and techniques.

AI can help managing many aspects when used appropriately, but we still need to define the right metrics for correct evaluation of the effects of our actions – individual and collective. Simulators cans help evaluating the situation, testing effect of influencing factors and consequence of decisions to take before doing. It applies in industry, city and in our lives.

The balance between the use of technology and human capacity should be preserved. Technology producers have tendency to produce software and devices that think instead of the human and take decisions for him/her. This kind of applications may replace human at the long run and reduce his cognitive capacity. That is why we have to design applications able to enhance human intelligence without switching it off.

Technology alone cannot save Planet, but can help us to be smarter.

References

1. Lenkowa, A.: Oskalpowana ziemia (1961)
2. Eckholm, E.: Losing ground. Environ. Sci. Policy Sustain. Dev. **18**(3), 6–11 (1976). https://doi.org/10.1080/00139157.1976.9930747
3. Cueto, J.: Planned Obsolescence – An Economic and Cultural Anxiety, July 2009. https://www.researchgate.net/
4. Mercier-Laurent, E.: The Innovation Biosphere –Planet and Brains in Digital Era. Wiley, Hoboken (2015)
5. WHO: Environment and Health. https://www.euro.who.int/en/health-topics/environment-and-health/Climate-change/data-and-statistics. Accessed 12 2021
6. OECD: Work in support of Climate actions, December 2019. https://www.oecd.org/environment/cc/OECD-work-in-support-of-climate-action.pdf
7. Changements climatiques: les éléments scientifiques du GIEC, August 2021. https://www.ecologie.gouv.fr/sites/default/files/21144_GIEC-2.pdf
8. IPCC 6th Assessment Report, March 2021. https://www.ipcc.ch/report/sixth-assessment-report-working-group-3/
9. Digital EU on twitter. https://twitter.com/DigitalEU
10. ISO 26000. https://www.iso.org/iso-26000-social-responsibility.html. Accessed 12 2021
11. https://new.siemens.com/global/en/company/topic-areas/digital-enterprise/discrete-industry.html. Accessed 7 2019
12. Future Factory Schneider Electric. https://www.youtube.com/watch?v=N2nbm5xHCjc
13. Schneider 2018–2019 Sustainability Report. https://www.schneider-electric.com/en/about-us/sustainability/
14. Carbon footprint of global trade, OECD (2015). https://www.itf-oecd.org/sites/default/files/docs/cop-pdf-06.pdf
15. https://www.dailymail.co.uk/sciencetech/article-9469541/Suez-canal-blockage-caused-spike-air-pollution.html. 14 Apr 2021
16. Waste sorting. https://www.futura-sciences.com/tech/actualites/robotique-robot-dope-ia-ameliorer-tri-dechets-73184/
17. Zhang, F., et al.: Toward an systemic navigation framework to integrate sustainable development into the company. J. Clean. Prod. **54**, 199–214 (2013). https://doi.org/10.1016/j.jclepro.2013.03.054
18. https://www.autodesk.com/industry/convergence. Accessed 12 2021
19. Dostatni, E., Diakun, J., Grajewski, D., Wichniarek, R., Karwasz, A.: Automation of the ecodesign process for industry 4.0. In: Burduk, A., Chlebus, E., Nowakowski, T., Tubis, A. (eds.) Intelligent Systems in Production Engineering and Maintenance, pp. 533–542. Springer International Publishing, Cham (2019). https://doi.org/10.1007/978-3-319-97490-3_51
20. Mercier-Laurent, E.: Innovation Ecosystems. Wiley, Hoboken (2011)
21. Mercier-Laurent, E.: Rôle de l'ordinateur dans le processus global de l'innovation à partir de connaissances. Université Jean Moulin, Lyon, HDR (2007)
22. https://sustainability.aboutamazon.com/environment/sustainable-operations/carbon-footprint. Accessed 12 2021
23. https://www.wonolo.com/

24. Capiluppi A, Baravalle A.: Matching demand and offer in on-line provision: A longitudinal study of monster.com, Web System Evolution, October 2010
25. Maher, M.L., Pu, P.: Issues and Applications of Case-based Reasoning to Design. Lavoisier, Paris (1997)
26. Nega-water. https://waterfamily.org/water-academie/
27. The carbon footprint of the Internet, April 2021. https://www.climatecare.org/resources/news/infographic-carbon-footprint-internet/
28. https://communication-responsable.ademe.fr/sites/default/files/guide-ademe-tic-quels-impacts.pdf
29. https://nsidc.org/. Accessed 4 2021
30. Kayakutlu, G., Mercier-Laurent, E.: Intelligence for Energy. In: Intelligence in Energy, pp. 79–116. Elsevier (2017). https://doi.org/10.1016/B978-1-78548-039-3.50003-1
31. Cavalucci, D.: TRIZ
32. Digital Europe. https://www.digitaleurope.org/policies/digital-transformation. Accessed 12 2021
33. Green Deal
34. https://www.oracle.com/cloud/digital-transformation.html#link9
35. MIT
36. Digital Strategy Europe. https://digital-strategy.ec.europa.eu/en/activities/digital-programme
37. Amidon, D.: The Innovation Strategy for the Knowledge Economy. Butterworth Heinemann, Oxford (1997)
38. Extreme programming. http://www.extremeprogramming.org/
39. Design Thinking
40. https://ec.europa.eu/info/funding-tenders/find-funding/eu-funding-programmes/digital-europe-programme_en
41. Ordonez de Pablos, P., Edvinsson, L. (Eds): Trends and Challenges for Intellectual Capital, chapter in Intellectual Capital in Organizations. Routledge, Milton Park (2015)
42. https://transformhealthcoalition.org/blogs/the-case-for-digital-health-report-launch-transform-health/
43. Mercier-Laurent, E.: Can AI efficiently support sustainable development. In: Artificial Intelligence for Knowledge Management, vol. 614. Springer AICT, Heidelberg (2021)
44. https://ec.europa.eu/info/strategy/priorities-2019-2024/european-green-deal_en
45. https://www.ecosmartphones.info
46. Schreiber, G., Wielinga, B., Breuker, J.: KADS – A Principled Approach to Knowledge-based System Development. Academic Press, Cambridge (1993)
47. Mercier-Laurent E.: L'approche Connaissance appliquée au retour d'expérience, EGC Lyon (2003)
48. Wong, A.: Ethics and regulation of artificial intelligence. In: Eunika Mercier-Laurent, M., Kayalica, Ö., Owoc, M.L. (eds.) AI4KM 2021. IAICT, vol. 614, pp. 1–18. Springer, Cham (2021). https://doi.org/10.1007/978-3-030-80847-1_1
49. Eliza. https://www.eclecticenergies.com/psyche/eliza
50. Top 36 chatbot applications https://research.aimultiple.com. 7 Apr 2021
51. Problem Solving Methods: Past, Present, and Future, AIEDAM, Spring (2008)
52. Cossentino, M., Kaisers, M., Tuyls, K., Weiss, G. (eds.): EUMAS 2011. LNCS (LNAI), vol. 7541. Springer, Heidelberg (2012). https://doi.org/10.1007/978-3-642-34799-3
53. Strous, L., Johnson, R., Grier, D.A., Swade, D. (Eds.): Unimagined Futures – ICT Opportunities and Challenges, Vol. 555, Springer AICT, Heidelberg (2020). https://doi.org/10.1007/978-3-030-64246-4
54. L'éthique ou Comment protéger l'IA de l'Humain pour protéger l'Humain de l'IA, Alsace Tech, 11 février 2021

55. https://intelligence.weforum.org/topics/a1Gb0000000pTDREA2?tab=publications
56. Nation States game. https://www.nationstates.net/
57. https://cordis.europa.eu/research-eu/en
58. United Nations Sustainability goals. http://www.gogeometry.com/mindmap/sdg-sustainable-development-goals-mind-map-10.jpg
59. Mercier-Laurent, E., Talens, G., Thivant, E.: Developing a knowledge base on climate change for metropolitan cities. In: Eunika Mercier-Laurent, M., Kayalica, Ö., Owoc, M.L. (eds.) Artificial Intelligence for Knowledge Management: 8th IFIP WG 12.6 International Workshop, AI4KM 2021, Held at IJCAI 2020, Yokohama, Japan, January 7–8, 2021, Revised Selected Papers, pp. 130–143. Springer International Publishing, Cham (2021). https://doi.org/10.1007/978-3-030-80847-1_9
60. Back market. https://www.backmarket.fr. Accessed Apr 2021
61. Repair café. https://www.repaircafe.org/fr/. Accessed Apr 2021
62. European Union – sustainable development. http://ec.europa.eu/environment. Accessed 7 2019
63. Michalski, R.S., Tecuci, G. (Eds): Machine Learning: A Multistrategy Approach, Volume IV. Morgan and Kaufmann, Burlington (1994). ISBN-13: 978–1558602519
64. Côte, F., Poirier-Magona, E., Perret, S., Roudier, P., Rapidel, B.,Thirion, M.-C.: The Agroecological Transition of Agricultural Systems in the Global South, Editions Quæ, June 2019
65. Mercier-Laurent, E., Edvinsson, L.: World Class Cooking for Solving Global: Reparadigming Societal Innovation. Emerald Publishing Limited (2021). https://doi.org/10.1108/978183867 1228
66. Ouskova Leonteva, A., Abdulkarimova, U., Wintermantel, T., Jeannin-Girardon, A., Parrend, P., Collet, P.: A quantum simulation algorithm for continuous optimization (poster). In: GECCO 2020: The Genetic and Evolutionary Computation Conference, Cancun, Mexico, p. 2, mars (2020)
67. https://www.youtube.com/watch?v=OEIeS12TcWU from min 4, Fedor (2019)

Crowdsourcing and Sharing Economic in the Smart City Concept. Influence of the Idea on Development and Urban Resources

Łukasz Przysucha[✉] [ID]

Wroclaw University of Economics and Business, Komandorska 118/120, 53-345 Wroclaw, Poland
lukasz.przysucha@ue.wroc.pl

Abstract. The sharing economy is a relatively new trend that occurs practically all over the world. It involves a complete change of organizational and distribution models. The main structure focuses on a distributed network of people and communities. This includes mutual service provision, sharing. This is definitely to use resources more efficiently. In recent years, we have seen a great popularity of systems based on the idea of the sharing economy. Crowdsourcing is a process related to the sharing economy in terms of obtaining information/knowledge. It is a process in which a given organization, a public entity, or a company performs outsourcing of tasks. The inquiry is usually directed to an unidentified, usually very wide group of people in the form of an open call. In the article, the author wonders if the sharing economy is the new economy? What is the impact of the sharing economy on Smart City? Is Crowdsourcing part of the sharing economy or is it the reverse? Will the sharing economy and crowdsourcing be the future of Smart City and what motivates residents to share together? There is a correlation between the development of the idea of the sharing economy and crowdsourcing and the development of Smart City. The article analyzes the relationship between these two trends and their impact on Smart City. The verification is carried out on the example of the city-state of Singapore, which has been occupying leading places in the Smart City rankings in recent years. An interesting aspect discussed in the article is also the exchange of knowledge in Smart City and the cycle of information between residents, decision-makers, and stakeholders and third parties, such as education or business.

Keywords: Sharing economy · Crowdsourcing · Smart city · Knowledge acquisition · Singapore

1 Introduction

From year to year, we observe an increasing development of globalization around the world. This is largely due to the mass media, the development of the Internet and electronics, as well as easier transport and popularization of travel.

© IFIP International Federation for Information Processing 2022
Published by Springer Nature Switzerland AG 2022
E. Mercier-Laurent and G. Kayakutlu (Eds.): AI4KMES 2021, IFIP AICT 637, pp. 19–31, 2022.
https://doi.org/10.1007/978-3-030-96592-1_2

Every now and then we learn about new trends in the world, as well as social development, which has an impact on the standardization of certain behaviors, especially in already developed countries. It can be observed that thanks to the facilitated means of communication and communication, it has become popular to share with the closest environment. In everyday life, we can see people streaming videos to friends and family, and photos presenting everyday activities, for example at home or work, while on the road, people willingly share cars, on the one hand caring for the environment, but also on the sociological side - the will to share information with other people as well as being together with them. We also travel and use shared services more and more often. Accommodation in hotels is as popular as platforms for spending the night together, for example renting rooms in one apartment with other, previously unknown people. Purchases of second-hand goods have also become popular, as well as the sharing of other people's goods for payment. People also use shared office spaces by renting them out by the hour. The sharing model has gained importance and is now expanding rapidly.

Sharing knowledge and information has become equally popular [1]. People want to inform the environment what they think about a given topic, what their beliefs and opinions are, and they want to influence the immediate environment around them.

As it can be seen, nowadays in every aspect of life, people share both their intellect and material goods, and this model is gaining popularity every year. The aspect of sharing from the Smart City perspective is also interesting. In this article, the author examines the impact of sharing trends on urban development and innovation in technology and intra-city communication, and tries to discuss how ideas will develop in the future. The phenomena of crowdsourcing and the sharing economy are analyzed on the example of Singapore, a city-state located in the center of Asia, a model Smart City, mentioned in many rankings.

2 Crowdsourcing in Smart City

Interaction between people is ubiquitous all over the world. Recently, it can be observed that more urbanized areas within agglomerations rely on innovative solutions that will develop better communication and information flow between their inhabitants. The attitude of the authorities is focused on the development of an intelligent society, i.e. one in which many factors interact [2–4]:

A. *Technologies*
   This is an area that supports intelligent society in terms of quick access to data, ensures data security and integrates many systems and tele-information planes in the city. It serves better communication and, indirectly, democracy in a given area through the possibility of publishing public data on, for example, administrative expenses and the possibility of aggregating these data by all residents in the city. Thanks to the latest technologies, it is also possible to analyze the past, Business Intelligence data support the city's development on the basis of archival data, and the authorities can estimate the future based on historical conclusions.

B. *Ecology*

Sustainable development and protection of natural resources and nature, it is necessary in many respects, mainly this aspect concerns ensuring good living conditions for future generations, but also adequate comfort for the current inhabitants of the city, e.g. anti-smog programs, appropriate fuel standards for cars moving in a given agglomeration training for young residents in the area of ecology, Smart Ecology solutions, installation of photovoltaic panels and heat pumps that do not affect the consumption of fossil fuels and do not produce carbon dioxide, there are emission-free.

C. *Sociology*

Coexistence in a very diverse society, full acceptance of various religions, sexual orientations, acceptance of a different worldview. Exchange of thoughts on neutral ground and with respect for other residents. Respect for tradition.

D. *Administration*

Intelligent management of society, interaction between residents, decision-makers and stakeholders such as universities, private companies. In Smart Society, it is necessary to obtain information from residents about their needs and beliefs, as well as to provide them with a partial possibility to decide about the immediate environment. Power is for the people and is involved in the development of the city, mainly with the residents in mind.

In the area of the aforementioned city administration, it is necessary to properly aggregate information from residents and stakeholders in order to know what are the expectations of other groups in the city. It comes to mind what tools can be effective in transmitting such data, what algorithms to use and how to collect them. Crowdsourcing is a process [5] in which, for example, a company, public institution, but also recently city administrations outsource tasks performed by employees to an unidentified, wide group of people in the form of an open call. A tool used by local administration can be modern means of communication such as social media, the Internet, mobile phones and others.

The discussed processes mainly take place in Smart Cities.

There are many definitions of Smart City depending on the perspective and aspects we consider. Some focus on defining a modern city as a whole of technology that favors the development of agglomeration, others on interpersonal aspects and human capital, which, when developed, gives its participants greater awareness of being in an intelligent society. Considering these definitions [6–8], it can be concluded that Smart City is a city that uses communication and information technologies [9] in order to increase the interactivity between users of the implemented systems, also increase the urban infrastructure, optimize processes between all areas in the agglomeration, synchronize city components, and also raising the awareness of residents of being an element of Smart Society, and thus activation in various aspects of life, e.g. active participation in the civic budget created by residents, impact on the immediate environment, as well as creating a profile of the resident and the consequences resulting from this fact.

Crowdsourcing is present in many cities in various forms. Recently, popular in Smart Cities is the civic budget (participatory budget), which takes into account the investment

propositions of residents and is aimed strictly at citizens. It consists in submitting proposals and investment needs in a given agglomeration. Residents directly decide what
they need and local authorities interact with them [10]. In this case, city decision-makers
only support the process from the technical side and verify the submitted proposals in
terms of compliance with the regulations (e.g. whether the tasks fall within the previously
indicated financial pool intended for all investments in the city). A simplified diagram
of the citizens' budget is presented below (Fig. 1).

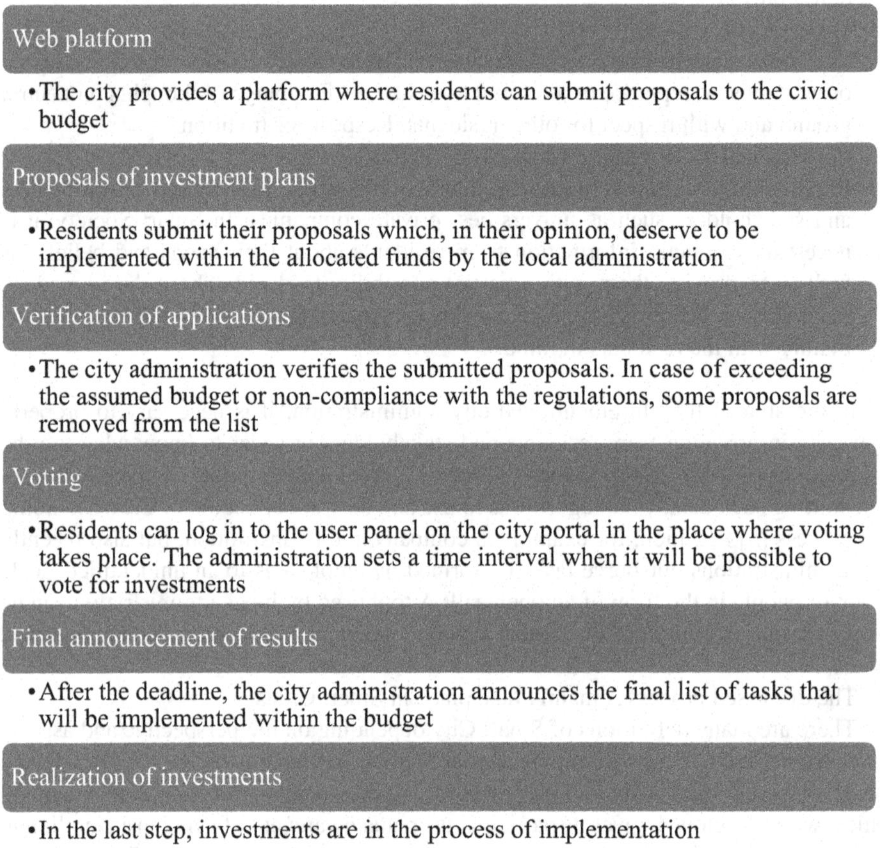

Web platform

- The city provides a platform where residents can submit proposals to the civic
  budget

Proposals of investment plans

- Residents submit their proposals which, in their opinion, deserve to be
  implemented within the allocated funds by the local administration

Verification of applications

- The city administration verifies the submitted proposals. In case of exceeding
  the assumed budget or non-compliance with the regulations, some proposals are
  removed from the list

Voting

- Residents can log in to the user panel on the city portal in the place where voting
  takes place. The administration sets a time interval when it will be possible to
  vote for investments

Final announcement of results

- After the deadline, the city administration announces the final list of tasks that
  will be implemented within the budget

Realization of investments

- In the last step, investments are in the process of implementation

Fig. 1. The mechanism of the civic budget

3 Sharing Economy

The sharing economy [11, 12] is primarily a social phenomenon involving a fundamental
change in both distributional and organizational models. It focuses mainly on the direct
provision of services between people, as well as sharing and co-creating many aspects
of life. Recently, there has been a lot of interest in this trend. In the past, this form of
cooperation referred only to the immediate family and circle of friends [13, 14] (Table 1).

Table 1. Advantages and disadvantages of sharing economy.

Advantages	Disadvantages
Increasing social ties between people - thanks to the sharing economy, a large part of society opens up to other people and meets new people. It definitely increases the sense of belonging from a psychological point of view	The threat to traditional business models, such as cars that are rented to residents to ultimately provide taxi services, may displace traditional taxi services in a given location
Positive impact on the environment - thanks to many solutions, incl. car sharing between cities or within agglomerations can reduce exhaust emissions, which translates into a cleaner environment	Too much data sharing can compromise privacy. It is uncertain whether the person who obtains the data is trusted and whether he will not use it for a wrong purpose
Effective management of available resources - by implementing some solutions associated with the sharing economy, it is possible to save time or money. Renting, for example, a bicycle or a car, optimizes reaching the destination, as well as does not require a financial contribution to the purchase of equipment	The lack of clear legislation on some aspects of the sharing economy means that a country can suffer less tax payments, and the processes are often unclear and inadequately defined
Increasing the possibility of finding a job - by providing services directly, you can bypass the monopolies of large companies and reduce the costs of providing services. This makes the data of the person more competitive	Lack of an unambiguous and independent reputation rating system for given individuals - people can often lose their entire career and business due to one negative comment in the Internet. All principles are based on social trust and represent an ideal model that does not assume many negative variants of behavior

Relationships between strangers are most often associated in the initial phase with a lack of trust, the need to obtain more information about the other person, and often protect themselves. Thanks to modern technologies, communication between users on both sides is much simpler and easier. Technologies overcome the obstacles before getting to know the other side, users can be in a safe area for them and communicate with others, transfer information and exchange knowledge. In the sharing economy, the field for sharing is much wider (Table 2).

Analyzing the above examples, one can see a great willingness to share people. There is a noticeable trend in the sharing economy. It is a socio-economic system built around the division of human and material resources. It includes the joint creation, production, distribution, trade and consumption of goods and services by different people as well as organizations.

On July 16–17, 2021, the Author conducted a study in the city of Wrocław, Poland in the field of the sharing economy and crowdsourcing. 200 residents in the city were asked. These were three open-ended questions. They are presented below with an illustration of the answers given (Fig. 2).

Table 2. Comparison of the most important systems based on the sharing economy.

System name	System description	Share analysis
Uber	An application supporting the use of carpooling. Thanks to the system, users can be taxi drivers and drive other people in their cars, as well as use other residents' journeys	At first, users were afraid to drive their cars with strangers, but the system quickly became popular. Thanks to the registration of trips and the general database of drivers, journeys have become safer than regular taxis and the application has gained a large crowd of fans
Airbnb	System supporting short-term rental of apartments and rooms. The system works all over the world, often recommended during vacation trips. Each resident can rent their house/flat. On the other side, there are strangers who are renting out places	People break the barrier of sleeping in someone's home, often with the owners in the other room. The system is becoming more and more popular. Another system that works in a similar system is Couchsurfing, which connects people who want to rent their home for free and tourists who want to stay without paying for the night
Wikipedia	Wikipedia is the largest encyclopedia on the Internet. It has 285 languages and nearly 300,000 editors. Each user can add their own content, and after moderation, it can be shared	The willingness to share knowledge is very high, especially in developed countries. Users very often use Wikipedia as the first source of finding knowledge, and search engine algorithms give it a significant advantage due to the large amount of original content
Ebay	Online shopping system. Used items can be sold. A very large proportion of Internet users use this system	Contrary to appearances, Internet users do not mind the fact that items are used or purchased as new from a foreign store without physically visiting it
Car Share Companies	Online shopping system. Used items can be sold. A very large proportion of Internet users use this system	Thanks to the application, the user can rent a car for minutes. Many smart cities have introduced this system to reduce the number of cars on the streets, and thus eliminate some exhaust fumes and pollution

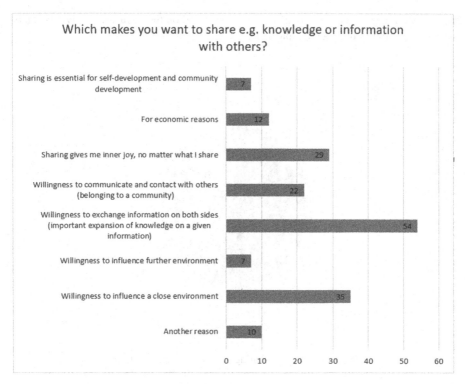

Fig. 2. Knowledge sharing survey – first question.

In the study, this question was asked to the inhabitants of the city of Wrocław to verify the reasons for wanting to share with others. Sharing can cover both tangible items as well as human capital, information and intangible items. The main aspect for which residents share is to make something available to get something. The questionnaire mainly provided answers in the field of information and knowledge exchange. Residents are curious about the information and knowledge possessed by other residents and by providing their resources they count on feedback. This is the most important reason for 27% of respondents. Another important aspect is the desire to influence the environment. Almost 18% want to influence the closer environment, while almost 4% want to influence further environment, which in total gives one in four respondents. A good example for smart cities are civic and participatory budgets, in which residents have a real say in matters of investment in the immediate vicinity as well as decisions made by local administration. In the case of Wrocław, according to data provided by the City Hall, 92,000 residents took part in 2020. Assuming that 80% of the inhabitants may be eligible to vote, for an urban population of 650,000, this is 520,000 inhabitants, which gives an interest of 17%, speaking of the local budget, which is consistent with the survey. The human factor is also an important element, which makes sharing between two or more people a joy. It is also consistent with wanting to belong to a group and community. A large proportion of the respondents marked this answer. The exchange also takes place for economic reasons, e.g. in the case of a desire to obtain material goods on the basis of barter exchange, participation in lotteries organized on the occasion of civic budgets aimed at mobilizing the society, and others (Fig. 3).

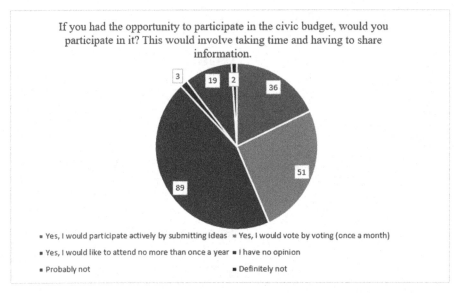

Fig. 3. Knowledge sharing survey – second question

Another aspect raised is the question of the willingness to participate in the civic budget. A very large proportion of the respondents were not aware of this form of decision-making in the field of local investments. Only every fourth respondent knew this mechanism. Ultimately, 88% of residents (176 people) expressed their willingness to participate in this initiative. It is interesting because in the previous question only 17% decided that the will to share is the most important for them. It can therefore be concluded that some residents put this goal in the next place for sharing knowledge and information, while others were encouraged and motivated only after presenting the entire voting platform. 18% of all respondents expressed a willingness to actively participate in the civic budget, and therefore a willingness to present their ideas to others. This group puts the issues of the local environment first. Over 44% of residents would like to participate in the project only once a year in a passive form by only voting for the best ideas in their opinion. On the other hand, 25% would like to participate once a month. Only 10% of the respondents are against participation in the citizenship project, which is a small percentage of the total.

Summing up, the civic budget enjoys great interest. Despite the fact that it is not well-publicized yet, many people participate both actively and passively in voting. The study shows that when the message is clear, participants overwhelmingly have an interest in participating in this initiative. People both share their knowledge, information and possessed material goods due to the hunger for knowledge of the other party, the desire to gain knowledge in a conversation as well as the desire to influence the close and further environment surrounding them. Many people share because they feel an inner belief that is not definable.

The sharing economy is a global trend and we are seeing a significant development in this aspect. Some scientists already define it as a type of economy. Interestingly, the sharing economy is present on many levels and everywhere in the world. It manifests itself in the will to share apartments and cars, but also knowledge and information. The latter can be associated with crowdsourcing, which supports the acquisition of knowledge from other people. The use of Crowdourcing is becoming more and more popular. The author draws attention to the dynamic development of crowdsourcing processes in the area of Smart City and agglomerations in developing and already developed countries.

4   Analysis of Crowdsourcing and the Sharing Economy Processes on the Example of Singapore

Singapore is a city-state located in Southeast Asia. Thanks to its location and history, it is one of the richest countries in the world and the financial center of the region. The city is also very developed in terms of being Smart. The most popular ranking positioning Smart City in the world - Smart City Index [15], created by the Institute for Management Development, has awarded Singapore the highest places in the ranking for many years (in the previous edition, Singapore, Helsinki and Zurich were the most intelligent cities). Singapore has been building an intelligent nation for many years, taking full advantage of the latest technology to improve citizens' lives, create more opportunities and build stronger communities. The city announced its strategy for the development of artificial intelligence: "Singapore National Artificial Intelligence Strategy" [16]. The document defines three strategic goals for the government:

1. Identify areas of focus and resources at the national level to find niches in which Singapore could become a world leader.
2. Develop the best model of cooperation between the government, companies and science in order to best realize the positive potential of artificial intelligence.
3. Identify the areas of "special concern" where, with the spread of artificial intelligence, change and risk management will be needed (eg labor market, decisions made by machines).

The entire strategy is developed in the time frame until 2030. This is another element in the implementation of the "Smart Nation" project. Smart Nation is an initiative of the Singapore government aimed at using information technology, networks and big data to create technological solutions. The key pillars supporting Singapore's Smart Nation goals are: digital economy, digital government and digital society.

1. Digital economy - The Singaporean government is trying to adapt digitization - its pace and size to the business activities of companies and the development of industry structures so that the work of companies is not disrupted and it does not adversely affect the development of society. Firms and businesses play a key role in the digital economy. The government wants excellent infrastructure and connectivity with

major Asian economies, as well as investment availability and well-developed tech-nology to help attract businesses and talent. They have a vision of building a dynamic economy that will remain attractive to foreign investment, with good opportunities for Singaporeans. Digitization of industries increases business efficiency and creates new jobs and opportunities.

2. Digital Government - The Digital Government Plan (DGB) was launched on June 5, 2018, setting out the vision and strategies for achieving Digital Government. 14 key performance indicators (KPIs) have been set to measure the progress of digitization. Policies and strategies, processes and organizational structures have also been introduced; talents were also recruited and trained to systematically use digital technologies and maintain the pace of development in the long term. Together, these efforts will get the government to digitization to the heart.

Below is a Smart Nation Framework diagram showing the implementation of a strategy for government digitization (Fig. 4).

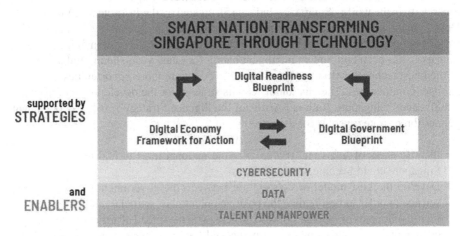

Fig. 4. Smart nation framework – Singapore. [17]

3. Digital Society - Digital inclusion aims to ensure that all Singaporeans have access to technology that can improve everyday life, and equip people with the skills and knowledge to use technology safely and confidently. To enable Singaporeans to maximize the opportunities of a digital society, improve their lives, connect with the world and have a level playing field for success, a Digital Readiness Plan has been issued.

The Government of Singapore is very involved in the development of digital society, but also Smart Nation, which is a key element for the further development of Smart City. Communication between public administration and residents, education and the business sector is extremely important. The government aggregates and processes data obtained from its stakeholders.

They are organizing crowdsourcing activities in the form of challenges such as idea generation contests, application development, and hackathons to address issues that Singaporeans may face.

All government agencies can effortlessly engage citizens and solicit ideas with Ideas !, Singapore's first government crowdsourcing portal. The portal, owned by the Chancellery of the Prime Minister (PMO) of Singapore, redefines citizen engagement by incorporating citizens' ideas and engaging with citizens to meet the nation's challenges. Ideas! is deployed on a government platform as a service (PaaS), also known as Nectar. Nectar was recently launched by the Singapore Government Technology Agency (GovTech) as a platform for hosting government digital services. With Nectar, government agencies can quickly develop, deploy, and scale applications. This application is a centralized platform for government administration activities. It enables all Singapore government agencies to use the collective wisdom of their citizens to aggregate a cost-effective way to generate and develop good ideas. Through the portal, each government agency can independently organize and manage crowdsourcing activities such as application development competitions, hackathons and campaigns. The audience can access the ideas! submit, comment and vote on the best ideas as well as share them on social networks.

Besides giving citizens a voice, Ideas! it encourages citizens to participate directly. Communities are empowered to lead the development process that shapes their lives. Ideas contributed by citizens can be selected as pilot projects for building citizenship and ownership (Fig. 5).

Path to citizen participation in cities

Fig. 5. Path to citizen participation in cities – Singapore. [17]

Singapore is a very developed and wealthy city, so it can invest and develop the subject of Smart City in its various aspects. The above-mentioned applications and platforms support the possibility of interaction between city decision-makers and residents. Users often use the mentioned programs for sharing and exchanging information/knowledge as well as material resources. The implemented ideas have the right resonance in the whole society and the government uses appropriate marketing to promote the implemented solutions. Properly prepared and implemented strategies are aimed at building a Smart Society, which is used, among others, by with the sharing economy and crowdsourcing as elements that expand and support its development.

5   Conclusions

From year to year, we observe an increasing development of civilization. The population is shifting from rural to urban areas. In connection with migration, problems arise in relation to communication, technological solutions and everyday life in large urban centers. Policy makers are informed about the needs of their residents and interact with them. Crowdsourcing presented in the article is an ideal tool for acquiring knowledge from residents and stakeholders. The author indicated the civic budget as an example of crowdsourcing. Attention should also be paid to a related phenomenon, sometimes referred to as the new economy - the sharing economy. It is thanks to him that we can observe the willingness of people around the world to share. These may be tangible goods, but the focus was on the analysis of information and knowledge transfer and sharing. The article shows that people want to share their knowledge in order to develop it and exchange news on a given topic. It is also important for them to develop the close and further environment that is directly related to them. On the example of cities and local communities, these can be civic (participatory) budgets, which are precisely a form of crowdsourcing. The article analyzes, on the example of Singapore, the mechanisms that auctioneer Smart Society and are closely related to crowdsourcing and sharing economy. The diagram below shows the role of crowdsourcing in the flow of knowledge (Fig. 6).

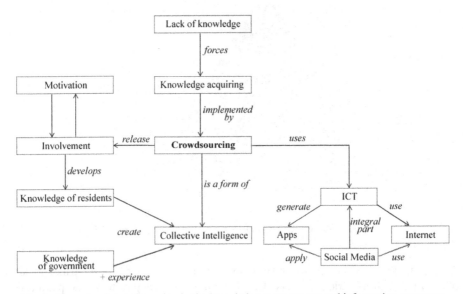

Fig. 6. The role of crowdsourcing in the knowledge management and information processes.

References

1. Owoc, M., Marciniak, K.: Usability of knowledge grid in smart city concepts. In: ICEIS, pp. 341–346 (2013)
2. Owoc, M.L., Weichbroth, P.: University students' research on artificial intelligence and knowledge management. a review and report of multi-case studies. In: Eunika Mercier-Laurent, M., Kayalica, Ö., Owoc, M.L. (eds.) AI4KM 2021. IAICT, vol. 614, pp. 66–81. Springer, Cham (2021). https://doi.org/10.1007/978-3-030-80847-1_5
3. Manda, M.I., Backhouse, J.: Towards a "Smart Society" through a connected and smart citizenry in South Africa: a review of the national broadband strategy and policy. In: Scholl, H.J., et al. (eds.) EGOVIS 2016. LNCS, vol. 9820, pp. 228–240. Springer, Cham (2016). https://doi.org/10.1007/978-3-319-44421-5_18
4. Albino, V., Berardi, U.: Smart cities: definitions, dimensions, performance, and initiatives. J. Urban Technol. **22**(1), 3–21 (2015)
5. Przysucha, Ł.: The concept of crowdsourcing in knowledge management in smart cities. In: Owoc, M.L., Pondel, M. (eds.) AI4KM 2019. IAICT, vol. 599, pp. 17–26. Springer, Cham (2021). https://doi.org/10.1007/978-3-030-85001-2_2
6. Dameri, R.P.: Smart city definition, goals and performance. In: Smart City Implementation. PI, pp. 1–22. Springer, Cham (2017). https://doi.org/10.1007/978-3-319-45766-6_1
7. Gołuchowski, J., Korzeb, M., Weichbroth, P.: Perspektywy wykorzystania architektury korporacyjnej w tworzeniu rozwiązań smart city. In: Roczniki Kolegium Analiz Ekonomicznych, pp. 85–98. Springer (2015)
8. Nam, T., Pardo, T.: Conceptualizing smart city with dimensions of technology, people, and institutions. In: Proceedings of the 12th Annual International Digital Government Research Conference: Digital Government Innovation in Challenging Times, pp. 282–291 (2011)
9. Domagała, P.: Internet of Things and Big Data technologies as an opportunity for organizations based on knowledge management. In: 2019 IEEE 10th International Conference on Mechanical and Intelligent Manufacturing Technologies (ICMIMT), South Africa (2019)
10. Adamiczka, H.: Civic budget as an instrument for meeting the needs of the citizens on the example of Wrocław. Architektura Krajobrazu **2**, 100–115 (2017)
11. Allen, D.: The sharing economy. Inst. Public Affairs Rev. Q. Rev. Polit. Public Affairs **67**, 24–27 (2015)
12. Hossain, M.: Sharing economy: a comprehensive literature review. Int. J. Hosp. Manage. **87**, 102470 (2020). https://doi.org/10.1016/j.ijhm.2020.102470
13. Mercier-Laurent, E.: Innovation Ecosystems. John Wiley & Sons, Hoboken (2011)
14. Özkuran, M.A., Kayakutlu, G.: Segmentation of social network users in Turkey. In: Mercier-Laurent, E., Boulanger, D. (eds.) AI4KM 2016. IAICT, vol. 518, pp. 119–131. Springer, Cham (2018). https://doi.org/10.1007/978-3-319-92928-6_8
15. Smart City Index by The Institute for Management Development
16. https://www.smartnation.gov.sg/why-Smart-Nation/NationalAIStrategy
17. https://www.tech.gov.sg/

Assessment of Smart Waste Management Systems with Spherical AHP Method

Mehmet Yörükoğlu$^{(\boxtimes)}$ ⓘ and Serhat Aydın ⓘ

National Defence University, Air Force Academy, Istanbul, Turkey
{myorukoglu1,saydin3}@hho.msu.edu.tr

Abstract. The increasing population and resource constraints in the world have made life in cities complex and difficult to manage. Especially the need for energy, water and other resources has become alarming, and the management and disposal of waste materials has also emerged as an important area of risk. Although waste management has been improved over time by methods such as recycling, it cannot fully solve the problems that have occurred. The need for landfill in cities, health and environmental problems have created the need for a more effective system to solve problems. It has become vital to create a sustainable life in cities that is more efficient, and which consumes less resources and improves the desired quality of life. The increase in these needs and especially developments in information technologies have created the "smart city" concept and applications that bring great innovations and facilities in many areas such as transportation, communication, health, security, energy efficiency, water use and waste management. Smart Waste Management Systems (SWMS) emerging in this context make an important contribution to the sustainability of today's cities and play an important role in carrying human life into the future. In addition, it increases the quality of life with cost optimization, energy efficiency, in-formation management, time saving, improving hygiene, and reduction in traffic, noise, bad odor and carbon emissions. This operation, defined as "smart", also sets an example for the use of artificial intelligence. The levels of the structures, processes and performances of the existing SWMSs show their success and thus the level of providing a sustainable life. The evaluation of these currently developing systems with their components will guide all relevant actors in the improvement of existing systems. Lack of a clear criterion used in the evaluation of SWMS required this assessment to be made with the MCDM (multi-criteria decision making) methods, which makes it possible to evaluate both objective and subjective criteria. As evaluating SWMS problem needs both tangible and intangible data, using fuzzy logic is a useful tool to solve the problem. Therefore, in the application section, the Spherical fuzzy AHP (Analytic Hierarchy Process) method is used to handle determined problem. In the application section, three alternatives are evaluated under four determined criteria. Results show that the problem is handled with Spherical fuzzy AHP method by efficiently and effectively.

Keywords: Sustainability · Smart city · Smart waste management system · Multi criteria decision making · Fuzzy sets theory · Spherical AHP method

© IFIP International Federation for Information Processing 2022
Published by Springer Nature Switzerland AG 2022
E. Mercier-Laurent and G. Kayakutlu (Eds.): AI4KMES 2021, IFIP AICT 637, pp. 32–43, 2022.
https://doi.org/10.1007/978-3-030-96592-1_3

1 Introduction

Today, the effects of smart systems that analyze digital data and reach accurate results are increasing rapidly. The analysis of real-time data with algorithmic techniques and engineering principles and their use in decision making create new and interesting research areas. Smart systems (SSs) include sensing, actuation, and control functions to identify and analyze a situation and make predictive or adaptive decisions based on available data, thereby taking intelligent actions [1]. SSs are software assets that perform a series of intelligent actions on behalf of a user or another program, which will involve making decisions using available data in a predictive or adaptive manner. SSs integrate the human mind with machine operations. They are dynamic and unstable, introducing various applications from health, aviation, logistics and ICT (information and communications technology). The improvement of SSs in many sectors can be associated with the reality that SS will reduce critical global problems such as climate change, health, and waste management. The improvement, downsizing and cheapening of the components used in SSs cause the increase in integration of SSs. Internet connections of the objects used in the SSs and users also create the combination of SSs [2].

Smart cities (SCs) are at the forefront of the areas where the mentioned combination is seen very clearly. A SC is an urban area that uses different types of electronic IoT (internet of things) sensors to collect data and then use the insights from this data to efficiently manage assets, resources, and services. SCs increase the quality of life of people with physical, digital, and human systems compatible with the environment, offering a modern, competitive, functional, and sustainable future. They are cities supported by advanced vital technologies. For a city to be called a SC, it is expected to have the following systems and parameters [3, 4]:

- Sufficient amount of water and energy,
- Effective solid waste management system,
- A sustainable informatics and communication infrastructure,
- An effective public transport and the mobile system of this public transport,
- E-municipal services suitable for the era,
- A successful, planned and programmed environmental management system,
- Age-appropriate crisis and emergency response system.

The concept of "sustainability", which was introduced through the report "Our Common Future" published by the World Commission on Environment and Development in 1987 [5], maintains its importance and even increases day by day, as can be seen with the "17 Sustainable Development Goals" put forward by the UN [6]. The following statement was used in the report: "Humanity; It has the ability to make development sustainable by meeting daily needs without compromising the ability of nature to respond to the needs of future generations." and the spirit of sustainability has been revealed. Although the first thing that comes to mind when it comes to sustainability is the protection of the environment, the concept of sustainability is actually a holistic approach that includes ecological, economic, and social dimensions [7].

Smart waste management system (SWMS) is an energy-efficient, economic, and technology-oriented system that optimizes the containers' and waste collection vehicles' capacities by minimizing the time loss. Thanks to the SWMS, end-to-end system control and citizen satisfaction are ensured, traffic congestion, working hours, fuel consumption and carbon emissions are reduced with waste collection optimization. It digitizes waste management, making it efficient and traceable. On the software side, smart waste audits are performed through online tools that use data analysis to review waste generation data, current waste collection practices and costs, and current onsite recycling levels. The SWMS includes decision support systems and performs process management and optimization, waste collection optimization, internet-based solutions, and data analysis. SWMS has a data management center; this center integrates garbage trucks, garbage containers, sensors, and their maps in a network. SWMS is designed to solve the current complex and costly process of inefficient routes served by a fleet of trucks at arbitrary and unconnected schedules [8, 9].

SWMSs draw attention in the transformation of cities into a sustainable living space. Through the implementation of IOT (internet of things), digitalization, and ICT (information and communication technology), achieving sustainability in waste management is becoming more possible, reliable, transparent, efficient, and optimum in the industrial revolution 4.0 era. IOT and ICT applications can help reduce the time and resources required to provide better performance of waste management toward sustainable and SSs [10]. Industrial Revolution 4.0 has brought significant changes in global waste management and value changes [11]. SWMS's framework consists of five aspects, including governance, economy, social, environmental, and technological [12].

The choices of the users determine the development direction and future features of SWMS, which causes the decisions about SWMS to emerge as a multi criteria decision making (MCDM) problem. Evaluation of SWMSs may contain conflicting criteria and therefore requires expert evaluations through linguistic variables. For this purpose, Spherical AHP Method, which is a newly introduced method as an extension of fuzzy sets, is used.

Spherical fuzzy sets (SFSs) are suggested by Kutlu Gündoğdu and Kahraman [13]. SFSs are based on the spherical fuzzy distances and satisfy the condition as follows $0 \leq \mu_{(x)}^2 + v_{(x)}^2 + \pi_{(x)}^2 \leq 1$. The hesitancy degree is represented by $\pi\_((x))$ and hesitancy degree can be determined in the spherical representation based on the given membership and non-membership values. So, a decision maker's hesitancy may be specified independently of membership degrees and non-membership degrees.

In this study, we used a hybrid model including SFSs and AHP method. The method, named Spherical Fuzzy AHP, is utilized to solve SWMS selection decision making problem. The criteria are determined to evaluate SWMS selection problem, and steps of the algorithm are applied to the problem.

In the literature with related studies, it is possible to see the waste management and some important applications of SWMS. Kayakutlu et al. evaluated the regional scenarios in the conversion of waste-to-energy with Bayesian network analysis [14] and used Value Stream Map in the detection and optimization of waste [15]. Yalçınkaya studied on a spatial modeling approach in organic waste management [16]. An IoT architecture was proposed for a proactive waste management by Aytaç and Korçak [17]. Kamm et al. presented a case study that represent a smart device implementation in smart waste collection processes [18]. Nizetic et al. focused on smart tecnologies in their study for promotion of energy efficiency, utilization of sustainable resources and waste management [19]. A mobile application using IoT in the waste collection system proposed by Kang et al., in this system it is aimed to dispose of the e-wastes generated at home by the users [20]. Rabbani et al. proposed a model which combine the decision-making techniques and location routing model for industrial waste management [21]. Sarkar and Sarkar [22] presented a study on the reduction of consumed energy and waste in a sustainable biofuel system. Marques et al. [23] presented a model on a multi-level IoT-based smart city infrastructure management architecture and evaluated the performance of this model with a real waste management problem. Sharma et al. [24] presented a study on the major IoT barriers faced in realizing smart city waste management. In this study, they revealed that internet connection problems, lack of standardization, policy norms, guidelines and regulations are the issues that prevent smart cities from showing better development, especially in terms of waste management. Aazam et al. proposed a cloud-based SWM mechanism in their study [25]. Zhang et al., listed the barriers of SWM in China and prioritized them with fuzzy DEMATEL (Decision-Making Trial and Evaluation Laboratory) [26]. Murugesan et al. proposed a theorical model of SWMS that uses WSN (Wireless Sensor Network) and IoT [27]. Abdallah et.al. presented a systematic research review that provides comprehensive analysis of the different AI models and techniques applied in SWM [10]. Sohag and Podder studied on an article where an IoT-based SWMS for sustainable urban life was presented [28]. Lu et al. presented an ICT based smart waste classification and collection system that is abstracted as a biobjective mathematical programming model to optimize the waste collection problem [29]. Esmaeilian et al. presented a review and concept paper about the future of waste management in smart and sustainable cities [30].

The rest of the paper is presented as follows: Steps of Spherical AHP are presented in Sect. 2, an application is given in Sect. 3, and Sensitivity Analysis is given in Sect. 4. Finally, conclusions are presented in Sect. 5.

2 Spherical AHP

This section covers the steps of the Spherical fuzzy AHP method.

Step 1. Build the hierarchical structure.

This step includes establishing a hierarchical structure. The hierarchical structure includes at least 3 levels; aim, which is at the top, attributes identified in the middle and alternatives at the bottom.

Step 2. Establish pairwise comparisons using spherical fuzzy judgment matrices.

Table 1 represents the linguistic terms with spherical numbers. After establishing the pairwise comparisons matrices, matrices need to be checked for consistency. For this purpose, linguistic terms are converted to their corresponding score indices, as seen in Table 1. After constructing pairwise comparison matrices, consistency formula, developed by Saaty [31], is applied.

Table 1. Linguistic scale

Linguistic expression	(μ, v, π)	Score index (SI)
Extremely preferred (ExP)	(0.9,0.1,0.0)	9
Very strongly preferred (VSP)	(0.8,0.2,0.1)	7
Strongly preferred (SP)	(0.7,0.3,0.2)	5
Moderately preferred (MP)	(0.6,0.4,0.3)	3
Equally preferred (EP)	(0.5,0.4,0.4)	1
Moderately low preferred (MLP)	(0.4,0.6,0.3)	1/3
Low preferred (LP)	(0.3,0.7,0.2)	1/5
Very low preferred (VLP)	(0.2,0.8,0.1)	1/7
Extremely low preferred (ELP)	(0.1,0.9,0.0)	1/9

Step 3. Compute the criteria and alternatives' local weights including spherical information.

In this step, Eq. (1) is utilized to get local weight of each alternative.

$$
\begin{aligned}
\mathrm{SWAM}_w(A_{S1}, \ldots\ldots, A_{Sn}) &= w_1 A_{S1} + w_2 A_{S2} + \ldots\ldots + w_n A_{Sn} \\
&= [1 - \prod_{i=1}^{n}(1 - \mu_{A_{Si}}^2)^{w_i}]^{\frac{1}{2}}, \prod_{i=1}^{n} v_{A_{Si}}^{w_i}, \\
[\prod_{i=1}^{n}(1 - \mu_{A_{Si}}^2)^{w_i} &- \prod_{i=1}^{n}(1 - \mu_{A_{Si}}^2 - \pi_{A_{Si}}^2)^{w_i}]^{1/2}
\end{aligned}
\tag{1}
$$

where $w = 1/n$

Step 4. Calculate the spherical global weights.

In this step, Eq. (2) is utilized to get Spherical fuzzy global weights.

$$
\prod_{j=1}^{n} \tilde{A}_{Sij} = \tilde{A}_{Si1} \otimes \tilde{A}_{Si2} \ldots \otimes \tilde{A}_{Sin} \forall i
$$

$$
i.e. \tilde{A}_{S11} \otimes \tilde{A}_{S12} = \left\langle \mu_{\tilde{A}_{S11}} \mu_{\tilde{A}_{S12}}, \left(v_{\tilde{A}_{S11}}^2 + v_{\tilde{A}_{S12}}^2 - v_{\tilde{A}_{S11}}^2 v_{\tilde{A}_{S12}}^2 \right)^{1/2}, \right.
$$

$$
\left. \left(\left(1 - v_{\tilde{A}_{S12}}^2\right) \pi_{\tilde{A}_{S11}}^2 + \left(1 - v_{\tilde{A}_{S11}}^2\right) \pi_{\tilde{A}_{S12}}^2 - \pi_{\tilde{A}_{S11}}^2 \pi_{\tilde{A}_{S12}}^2 \right)^{1/2} \right\rangle
\tag{2}
$$

Then, the final score of each alternative is determined via Eq. (3).

$$
\tilde{F} = \sum_{j=1}^{n} \tilde{A}_{Sij} = \tilde{A}_{Si1} \oplus \tilde{A}_{Si2} \ldots \oplus \tilde{A}_{Sin} \forall i
$$

$$
i.e. \tilde{A}_{S11} \oplus \tilde{A}_{S12} = \left\langle \left(\mu_{\tilde{A}_{S11}}^2 + \mu_{\tilde{A}_{S12}}^2 - \mu_{\tilde{A}_{S11}}^2 \mu_{\tilde{A}_{S12}}^2 \right)^{1/2}, v_{\tilde{A}_{S11}} v_{\tilde{A}_{S12}}, \right.
$$

$$
\left. \left(\left(1 - \mu_{\tilde{A}_{S12}}^2\right) \pi_{\tilde{A}_{S11}}^2 + \left(1 - \mu_{\tilde{A}_{S11}}^2\right) \pi_{\tilde{A}_{S12}}^2 - \pi_{\tilde{A}_{S11}}^2 \pi_{\tilde{A}_{S12}}^2 \right)^{1/2} \right\rangle
\tag{3}
$$

Step 5. The final score of each alternative is defuzzified by using the score function by Eq. (4).

$$S(\tilde{w}_j^s) = \sqrt{\left| 100 * \left[\left(3\mu_{\tilde{A}_s} - \frac{\pi_{\tilde{A}_s}}{2} \right)^2 - \left(\pi_{\tilde{A}_s} - \frac{v_{\tilde{A}_s}}{2} \right)^2 \right] \right|} \qquad (4)$$

Step 6. At the last step, the alternatives are ordered according to their defuzzified final scores in descending order.

3 Application

In this part of the paper, the SWMS selection problem is handled by spherical AHP method. To the this aim, first four different alternatives are determined as follows; A_1, A_2, A_3 and A_4. After a literature review, four criteria have been determined. Criteria are Governance (C_1), Social (C_2), Environmental (C_3), and Technological (C_4). Pairwise comparison matrices are fulfilled by us after gathering data from different SWMS users' experience.

Step 1. The hierarchical structure is established as seen in Fig. 1.

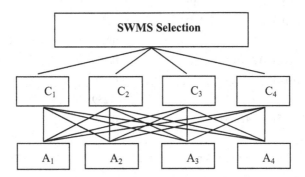

Fig. 1. The hierarchical structure

Step 2. Pairwise comparisons are established as seen in Tables 2, 3, 4, 5 and 6. Tables 2, 3, 4, 5 and 6 also includes spherical weights ($\tilde{w}^s$) and crisp weights ($\overline{w}^s$), and consistency ratios (CR).
Step 3. Spherical fuzzy local weights of alternatives are calculated as seen Table 7.
Step 4 and Step 5. Spherical fuzzy global preference weights and defuzzified values of final scores of each alternative can be seen in Table 8.
Step 6. Alternatives are ranked in descending order according to the defuzzified final scores.

Alternative 1 > Alternative 4 > Alternative 3 > Alternative 2.

"Alternative 1" is chosen the best according to the ranking scores.

Table 2. Pairwise comparison of criteria

Criteria	C_1	C_2	C_3	C_4
C_1	EP	SP	VSP	SP
C_2	LP	EP	MP	EP
C_3	VLP	MLP	EP	MLP
C_4	LP	EI	MP	EP
	$\tilde{w}^s$		$\overline{w}^s$	CR
C_1	(0.70, 0.29, 0.22)		0.352	0.027
C_2	(0.49, 0.46, 0.34)		0.232	
C_3	(0.39, 0.58, 0.30)		0.184	
C_4	(0.49, 0.46, 0.34)		0.232	

Table 3. Pairwise comparison of alternatives according to C1

C_1	A_1	A_2	A_3	A_4
A_1	EP	SP	MP	MP
A_2	LP	EP	MLP	MLP
A_3	MLP	MP	EP	LP
A_4	MLP	MP	PP	EP
	$\tilde{w}^s$		$\overline{w}^s$	CR
A_1	(0.61, 0.37, 0.30)		0.298	0.029
A_2	(0.41, 0.56, 0.31)		0.190	
A_3	(0.51, 0.44, 0.35)		0.239	
A_4	(0.51, 0.44, 0.35)		0.239	

Table 4. Pairwise comparison of alternatives according to C2

C_2	A_1	A_2	A_3	A_4
A_1	EP	MP	VSP	SP
A_2	MLP	EP	MP	MP
A_3	VLP	MLP	EP	MLP
A_4	LP	MLP	MP	EP
	$\tilde{w}^s$		$\overline{w}^s$	CR
A_1	(0.67, 0.31, 0.25)		0.338	0.052
A_2	(0.57, 0.41, 0.30)		0.278	
A_3	(0.39, 0.58, 0.30)		0.184	
A_4	(0.45, 0.53, 0.30)		0.216	

Table 5. Pairwise comparison of alternatives according to C3

C_3	A_1	A_2	A_3	A_4
A_1	EP	MP	VLP	MLP
A_2	MLP	EP	ELP	LP
A_3	VSP	ExP	EP	MP
A_4	MP	SP	MLP	EP
	$\tilde{w}^s$		$\overline{w}^s$	CR
A_1	(0.46, 0.53, 0.30)		0.218	0.033
A_2	(0.36, 0.62, 0.28)		0.169	
A_3	(0.76, 0.24, 0.20)		0.387	
A_4	(0.57, 0.41, 0.30)		0.278	

Table 6. Pairwise comparison of alternatives according to C4

C_4	A_1	A_2	A_3	A_4
A_1	EP	MP	SP	EP
A_2	MLP	EP	MP	LP
A_3	LP	MLP	EP	LP
A_4	EP	SP	SP	EP
	$\tilde{w}^s$		$\overline{w}^s$	CR
A_1	(0.59, 0.37, 0.32)		0.284	0.043
A_2	(0.47, 0.51, 0.31)		0.223	
A_3	(0.39, 0.59, 0.29)		0.181	
A_4	(0.62, 0.35, 0.30)		0.302	

Table 7. Spherical fuzzy weighted matrix

Alternatives	C_1	C_2
A_1	(0.42, 0.46, 0.35)	(0.37, 0.51, 0.37)
A_2	(0.28, 0.61, 0.35)	(0.28, 0.59, 0.40)
A_3	(0.35, 0.51, 0.39)	(0.19, 0.70, 0.37)
A_4	(0.35, 0.51, 0.39)	(0.22, 0.67, 0.37)
	C_3	C_4
A_1	(0.19, 0.71, 0.36)	(0.33, 0.54, 0.39)
A_2	(0.15, 0.76, 0.33)	(0.22, 0.66, 0.38)
A_3	(0.27, 0.64, 0.35)	(0.18, 0.71, 0.36)
A_4	(0.23, 0.67, 0.36)	(0.30, 0.57, 0.39)

Table 8. Defuzzified final score values and ranking of alternatives

	Total	Total Score	Ranking
A_1	(0.63, 0.09, 0.54)	9.679	1
A_2	(0.46, 0.18, 0.60)	6.128	4
A_3	(0.49, 0.16, 0.59)	6.759	3
A_4	(0.53, 0.13, 0.59)	7.408	2

4 Sensitivity Analysis

In this section, a sensitivity analysis is performed. Different weights are assigned to criteria and are analyzed to observe how much it would influence the final scores of alternatives. In the first case, the criteria weights are 0.352, 0.232, 0.184, and 0. 232 respectively and the final scores are obtained as 9.679, 6.128, 6.759, 7.408 respectively.

In the second case, the criteria weights are 0.10, 0.70, 0.10, and 0.70 respectively and the final scores are obtained as 6.345, 5.254, 7.009, 7.408 respectively. Therefore, the fourth alternative is selected best alternative, and the third alternative is chosen the second alternative. As a result, we can say that the second criterion has big impact on third and fourth alternative.

In the third case, the criteria weights are 0.70, 0.10, 0.10, and 0.70 respectively and the final scores are obtained as 12.345, 5.254, 6.009, 2.208 respectively. Therefore, the first alternative is selected best alternative. Briefly, we can say that the first criterion has a big impact on first alternative.

In the fourth case, the criteria weights are 0.10, 0.10, 0.70, and 0.70 respectively and the final scores are obtained as 6.345, 4.254, 12.009, 7.208 respectively. Therefore, the third alternative is selected best alternative. Briefly, we can say that the third criterion has a big impact on third alternative.

In the last case, the criteria weights are 0.10, 0.10, 0.10, and 0.70 respectively and the final scores are obtained as 10.345, 8.254, 8.009, 13.208 respectively. Therefore, the fourth alternative is selected best alternative. Briefly, we can say that the fourth criterion has a big impact on fourth alternative.

According to the sensitivity analysis results we can say that the best alternative decision is sensitive to the changes in the criteria weights. Figure 2 illustrates the results of sensitivity analysis.

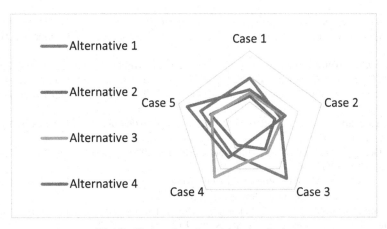

Fig. 2. The results of sensitivity analysis.

5 Conclusion

With the development of ICT, IoT applications and artificial intelligence, smart systems have become widespread in all areas of life. The transformation of cities, which are one of the basic platforms of life, into smart cities is progressing and smart applications about cities are getting preferable. In this process, SWMS stands out as one of the systems that continue to develop in the city and affect human life in a significant way.

The development of SWMS still goes on, this formation is guided by the development of information technologies and especially by the preferences of the users. User preferences determine the future features of SWMS, which continues to evolve. Evaluation of SWMSs may involve different conflicting criteria. Since SWMS assessment is considered as an MCDM problem, there is a need for this assessment to be obtained through linguistic variables and experts' assessments. In this study, a newly developed method, spherical AHP method, as an extension of fuzzy set, is used to get the opinions of experts.

In the application section, we evaluated different smart waste management systems. Therefore, we determined four different alternatives and four different crtieria. We used newly developed method named spherical fuzzy AHP to determine the best alternative. Finally, we ranked the alternatives in descending order according to the used methods result.

In future, the smart waste management system evalution problem can be solved by different extension of fuzzy set such as intuitionistic fuzzy TOPSIS, Pythagorean fuzzy TOPSIS, hesitant fuzzy TOPSIS, etc.

References

1. Lim, C., Maglio, P.P.: Data-driven understanding of smart service systems through text mining. Serv. Sci. **10**(2), 154–180 (2018)

2. Tränkler, H.R., Kanoun, O.: Smart systems and devices: Innovative key modules for engineering applications. In: Conference Smart Systems and Devices, pp. 3–12, Hammamet, Tunisia, (2001)
3. Smart Cites World Homepage. https://www.smartcitiesworld.net. Accessed 26 July 2021
4. UN ITU-T Focus Group on Smart Sustainable Cities: An overview of smart sustainable cities and the role of information and communication technologies. Focus Group Technical report (2014)
5. UN Secretary-General: Report of the World Commission on Environment and De-velopment, A/42/427 (1987)
6. UN General Assembly: Transforming our world: The 2030 Agenda for Sustainable Development, A/RES/70/1 (2015)
7. Farrell, A., Hart, M.: What does sustainability really mean? The search for useful indicators. Environ. Sci. Policy. Sustain. Dev. **40**(9), 4–31 (1998)
8. Hassan, S.S., Jameel, N.G.M., Şekeroğlu, B.: Smart solid waste monitoring and collection system. Int. J. Adv. Res. Comput. Sci. Softw. Eng. **6**(10), 7–12 (2016)
9. Kamm, M.R., Gau, M., Schneider, J., Brocke, J.: Smart waste collection processes, a case study about smart device implementation. In: Proceedings of the 53rd Hawaii International Conference on System Sciences, pp. 6619–6628 (2020)
10. Abdallah, M., Talib, M.A., Feroz, S., Nasir, Q., Abdalla, H., Mahfood, B.: Artificial intelligence applications in solid waste management: a systematic research review. Waste Manage. **109**, 231–246 (2020)
11. Zhang, A., Venkatesh, V.G., Wang, J.X., Venkatesh, M., Wan, M., Qu, T.: Drivers of industry 4.0 enabled smart waste management in the supply chain operations. Production Planning and Control. (Accepted/In press)
12. Fatimah, Y.A., Govindan, K., Muriningsih, R., Setiawan, A.: Industry 4.0 based sustainable circular economy approach for smart waste management system to achieve sustainable development goals: a case study of Indonesia. J. Cleaner Prod. **269**, 122263 (2020). https://doi.org/10.1016/j.jclepro.2020.122263
13. Gündoğdu, F.K., Kahraman, C.: Spherical fuzzy sets and spherical fuzzy TOPSIS method. J. Intell. Fuzzy Syst. **36**(1), 337–352 (2019)
14. Kayakutlu, G., Daim, T., Atlay, M., Yulianto, S.: Scenarios for regional waste management. Renew. Sustain. Energy. Rev. **74**, 1323–1335 (2017)
15. Kayakutlu, G., Şatoğlu, Ş.I., Durmuşoğlu, B.: Waste detection and optimization by applying Bayesian causal map technique on value stream maps. In: Proceedings of the 19th International Conference on Production Research (2007)
16. Yalcinkaya, S.: A spatial modeling approach for siting, sizing and economic assessment of centralized biogas plants in organic waste management. J. Clean. Prod. **255**, 120040 (2020). https://doi.org/10.1016/j.jclepro.2020.120040
17. Aytaç, K., Korçak, Ö.: IoT based intelligence for proactive waste management in quick service restaurants. J. Clean. Prod. **284**, 125401 (2021)
18. Kamm, M.R., Gau, M., Schneider, J., Brocke, J.V.: Smart waste collection processes, a case study about smart device implementation. In: Proceedings of the 53rd Hawaii International Conference on System Sciences (2020)
19. Nizetic, S., Djilali, N., Papadopoulos, A., Rodrigues, J.P.C.: Smart technologies for promotion of energy efficiency, utilization of sustainable resources and waste management. J. Clean. Prod. **231**, 565–591 (2019)
20. Kang, K.D., Kang, H., Ilankoon, I.M.S.K., Chong, C.Y.: Electronic waste collection systems using Internet of Things (IoT): household electronic waste management in Malaysia. J. Clean. Prod. **252**, 119801 (2020)

21. Rabbani, M., Sadati, S.A., Farrokhi-Asl, H.: Incorporating location routing model and decision making techniques in industrial waste management: application in the automotive industry. Comput. Indust. Eng. **148**, 106692 (2020)
22. Sarkar, M., Sarkar, B.: How does an industry reduce waste and consumed energy within a multi-stage smart sustainable biofuel production system? J. Clean. Prod. **262**, 121200 (2020)
23. Marques, P., et al.: An IoT-based smart cities infrastructure architecture applied to a waste management scenario. Ad Hoc. Netw. **87**, 200–208 (2019)
24. Sharma, M., Joshi, S., Kannan, D., Govindan, K., Singh, R., Purohit, H.C.: Internet of Things (IoT) adoption barriers of smart cities' waste management: an Indian context. J. Clean. Prod. **270**, 122047 (2020)
25. Aazam, M., St-Hilaire, M., Lung, C., Lambadaris, I.: Cloud-based smart waste management for smart cities. In: IEEE 21st International Workshop on Computer Aided Modelling and Design of Communication Links and Networks (CAMAD), pp. 188–193 (2016)
26. Zhang, A., Venkatesh, V.G., Liu, Y., Wan, M., Qu, T., Huisingh, D.: Barriers to smart waste management for a circular economy in China. J. Clean. Prod. **240**, 1181–1198 (2019)
27. Murugesan, S., Ramalingam, S., Kanimozhi, P.: Theoretical modelling and fabrication of smart waste management system for clean environment using WSN and IOT. Proc. Mater. Today **45**(2), 1908–1913 (2021)
28. Sohag, M.U., Podder, A.K.: Smart garbage management system for a sustainable urban life: an IoT based application. Internet of Things **11**, 100255 (2020)
29. Lu, X., Pu, X., Han, X.: Sustainable smart waste classification and collection system: a bi-objective modeling and optimization approach. J. Clean. Prod. **276**, 124183 (2020)
30. Esmaeilian, B., Wang, B., Lewis, K., Duarte, F., Ratti, C., Behdad, S.: The future of waste ma-nagement in smart and sustainable cities: a review and concept paper. Waste Manage. **81**, 177–195 (2018)
31. Saaty, T.L.: The Analytic Hierarchy Process, pp. 78–99. McGraw- Hill, New York (1980)

Zero Carbon Energy Transition in the Kitchens

Atilla Kılınç[1]([✉]) [iD] and M. Özgür Kayalica[2] [iD]

[1] Istanbul Technical University, 34469 Istanbul, Turkey
kilinca20@itu.edu.tr
[2] Institute of Energy at Istanbul Technical University, 34469 Istanbul, Turkey
kayalica@itu.edu.tr

Abstract. The United States (US) is ranked as the second country with the highest carbon emission after China. The transformation to improve energy efficiency in the US has a critical global impact on carbon emission reduction. Therefore, any attempt towards transformation will count. Designing a new optimization model for the transformation of the kitchens is no exception and could be seen as innovative and realistic. This study aims to combine carbon emission reduction and energy efficiency by using a carbon tax system within the jurisdiction of local authorities to transform cooktop ovens in kitchens in the South Atlantic region. The South Atlantic census region is selected for the analysis due to its high propane usage. The carbon emissions are reduced by 1.2% with the proposed optimization model using the RECS (Residential Consumption Survey) data set. Based on the benefits of the first application, a new model is developed to look at the future with increasing demand. A regression-based machine learning is used in the R software to create a general model that predicts the efficiency increases. The model is constructed to assume that 100% of the propane cooktop ovens are converted into electric induction cooktop ovens. The proposed model will have two positive results. First, it will encourage the replacement of propane cooking devices with energy-efficient electric induction cooktop ovens to reduce carbon emissions. Second, the energy accessibility will be increased as energy-efficient appliances will be donated to the users by using the budget created through the carbon tax incomes.

Keywords: Energy transition · Energy efficiency · Carbon tax · Regression by R

1 Introduction

1.1 Background

In the 21st century, the rapid population growth, economic changes, and differentiation in living spaces with technological developments have led to increased energy consumption around the Globe. With this increase, countries have taken the energy demand function as a basis while determining the policies and designing the management systems with regulations. As a result, the installed power capacities are improved, energy markets are restructured, and the energy systems are considered whole. Moreover, the excessive

© IFIP International Federation for Information Processing 2022
Published by Springer Nature Switzerland AG 2022
E. Mercier-Laurent and G. Kayakutlu (Eds.): AI4KMES 2021, IFIP AICT 637, pp. 44–62, 2022.
https://doi.org/10.1007/978-3-030-96592-1_4

increase in demand caused the diversification of energy sources and created complex systems in the energy markets. Energy efficiency can be counted as a primary component in forecasting the future of the energy sector under different scenarios. Implementing policies such as carbon tax would be beneficial in accelerating the energy transformation will help us set up an optimization strategy in this model. The most innovative part of the model is covering the entire campaign with the carbon tax that will come into operation. Thus, it will not be an economic burden on the state budget, and a step will be taken in social responsibility by donating these products to the public.

1.2 Research Objective

The main aim of this optimization model is to encourage users who cook with propane in their kitchens to use energy-efficient electric induction cooktops, thereby reducing carbon emissions in the United States and contributing to the 2050 targets in the context of the Paris agreement. The sizeable geographical surface and many different energy sources provided a positive external effect on the optimization conditions of the energy system in the US. However, the critical problem has been experienced in consumption behaviors and product habits. The first is the classification of the regions where the households are located. Living spaces are divided into three categories as Urban Cluster, Urban, and Rural have caused a change in energy consumption approaches.

Secondly, after the attacks on September 11, 2001, both the world economic crisis and the global or international security crisis, the escape from the cities in the US caused urban clusters and rural areas to change. With this unexpected change, people from all classes began to live in the suburbs, which caused the household areas to expand and grow and the energy demand to increase with a linear function. By increasing the number of households, the number and size of the houses have changed due to this expansion trend, and behavior patterns in the energy demand function have emerged. According to the South Atlantic region's RECS data, which is considered for this optimization model, the results close to each other indicate how the region was affected by this migration movement.

This article is organized as follows. In Sect. 2, the recent literature related to carbon tax application in the United States will be shown as clusters. Section 3 will contain the methodological approaches which try to create the optimization model. Thus, the model's main structure will be shaped through the method review. In Sect. 4, the model will be presented together with the application. Finally, we conclude with the results of the model application.

2 Literature Review

In general, the increasing sensitivity to reducing carbon emissions due to both the Kyoto Protocol and the Paris Agreement has started to show its effect in the United States. Therefore, more smart homes, energy efficiency certificates like LEED, and micro-grids created with solar power-based energy systems and batteries are being implemented rapidly [1].

Numerous articles can be found in the literature on carbon tax implementation; however, this article has selected the most relevant studies for review. Firstly, similar articles generally focus on how the carbon tax will function in the context of climate change and what kind of policy should be followed in terms of its limitation. Therefore, it contributed to how the carbon tax will affect the optimization model in this scenario. Two studies have been sufficient justification to display how the carbon tax as a social policy element created in this project will have a transformative effect. Although the carbon tax is thought to be a regressive system, since energy consumption takes a significant amount in the general income of the middle and lower-middle-income groups, it is clear that a tax that is likely to be applied in energy will cause a decrease in this income order [2, 3].

Secondly, there would be a classification of the consequences of implementing the carbon tax based on consumption trends and its general effects on households' income in the United States. The first article evaluates the different effects of carbon taxes applied within specific rates between household income groups. At this point, the focus is on both the source side and the user side of energy. The question is how the carbon tax functions through the integration between these two are justified? This integration is directly related to the optimization model of the article [4]. In other works, state that if not implemented very carefully, carbon taxes negatively affect investment supports and incentives. Therefore, the optimization model in this article is based on households only. In addition, the harmful conditions mentioned in these papers are neutralized since all the income generated by the carbon tax is returned to the consumers [5, 6].

Thirdly, to see the effect of the carbon tax policy would be so valuable to implement in this model. Therefore, similar articles showed the positive and negative impact of the tax policy. The first article focuses on designing a carbon tax that is both low-cost and highly effective, based on carbon tax practices in the international arena. Another issue discussed here is explaining what support mechanisms should be used to make carbon taxes more positive to reduce carbon emissions effectively. Therefore, it is said that the economic situation should be examined when applying a carbon tax [7]. The other four similar articles focus on how carbon taxes should be used in the context of climate change. The main aim of these articles is global warming and the relationship of the economic problems caused by this problem with the carbon tax. Considering optimization-centered, it can be said that it is directly related to the approach put forward in this model. In addition, the increase in the constructions based on the possibility of carbon taxes turning into emission fees, in the long run, emerges as new options to guide energy consumption trends. It is seen that such practices, which are included in the system to achieve the targets for 2030 in the context of both the United States and the European Union, will also be transferred to emerging markets soon [8–12]. These similar articles concern the sources that refer to the Paris Agreement. As noted throughout this modeling article, the United States is second only to The Republic of China for carbon emissions. However, the most crucial difference here is that the United States citizens cause the most carbon emissions, according to the population numbers.

Therefore, many of the articles refer to this problem and give examples of measures to be taken. The most important of these examples are seen as the carbon tax. Such tax-based practices are requested to be put on the agenda quickly to ensure environmental justice, change the public's choices behaviorally, and raise awareness about global warming and climate change. More importantly, it is questioned whether the public is ready to pay such a tax if such an approach occurs. For this reason, the carbon tax is seen not only as an income element but also as a vital state instrument in which both the public's declaration of responsibility and government policies are brought to a reconciliation point [13–15].

As final, some publications deal with all aspects of the carbon tax system. These publications offer a wide range of information, starting from carbon tax as a method to future carbon tax policies. These articles created a base review to express the optimization model in this article. The first article has the approaches that suggest that carbon will be seen as a tax element and a parameter to set prices in energy [16]. It is seen that the policy approach that proposes that the analysis of the carbon tax will be applied with different methods has developed. Therefore, the importance of carbon tax analysis methods should be emphasized [17]. The three articles need to be treated as a group within a group. In this group, the relationship between a carbon tax and macroeconomic balances is discussed concerning global warming and achieving the targets of climate protocols. It also focuses on the technology perspective of this situation. It should be said that this group, which is very important in every respect, has a severe theoretical impact on this article [18–20]. The articles that follow are the articles that predict that the carbon tax is now a severe measure and is starting to change the economy [21]. What stands out here is the fact that carbon, even as a tax type, has shown its impact on the future with the effect it has had from the time it first emerged is consistency [22–24].

As we mentioned before, although there are many articles written on similar topics, almost none of them match the qualifications of this research. There are similarities of various points. Starting from these points can be seen as the most logical. The aim is to produce a new model after seeing a few different models working on their research efficiency. For this reason, taking advantage of the options offered by Excel Solver and R Studio and designing the framework of the model according to RECS data will enable us to include the components for this future optimization scenario effectively into the model.

3 Methodology

The methodological application in this research first started with editing the Residential Energy Consumption Survey data obtained from the Energy Information Administration. After the conditional probabilities created by ethical and legal restrictions regarding the region's characteristics, the data began to be processed. Various modeling methods were examined in the process after interacting with the RECS data. Later, it was thought that there was a possibility of optimization with LEAP or OseMoSYS. However, starting from the idea of the first carbon tax, there was a consensus that regional application would be a more logical way to see the efficiency of the tax. Therefore, first MARKAL, TIMES, LEAP, and OseMoSYS method reviews have been added to the article. When the first optimization examples were built with Excel Solver, clear solutions were obtained.

Then, a detailed LEAP model review is included in the article and the development staging of the model. It would be logical to stay in this section to understand why the constructed carbon tax model is not included in such optimization programs.

The Excel solver program and the R statistics program used in the Zero Carbon net article provided all the necessary conditions for optimization. The most crucial point here is the cluster method within the RECS data itself. Thanks to Excel Solver and R, the number of rows of the data has decreased. A clear optimization structure has emerged with the solutions of programs such as Excel Solver and R more effectively in a shorter time. In the datasets taken as the basis to predict future energy demands using the LEAP model, the population structure of the region or regions, energy efficiency levels, the growth curve of the economy, industrial production structure, and all other external factors related to energy are taken as parameters [25].

In this example, where LEAP will be applied over the United States data set, it should be thought that it will produce a discourse outside the existing literature regarding being an analysis other than the current energy models. With the introduction of carbon taxes, which emerged primarily in the context of the trade wars between the United States and the Republic of China, and it will find application in the coming terms due to the reduction of carbon emissions, many energy market models need to update their formulations as seen as a possible. For this reason, to foresee that it would be a more logical maneuver to use LEAP when doing forward modeling [26]. Since the beginning of the optimization model, we have experienced different programs and developed different formulations to catch the most suitable and effective model design. Therefore, the article's content aims to prove how the excel solver and R studio work collaboratively to create a peculiar carbon tax optimization related to South Atlantic Region in the United States.

4 Modelling Approach

4.1 Introduction

The United States comes second in the countries that provide the most carbon emissions in the world. Therefore, the economic size created by the energy transformation of the system in the energy efficiency center seems to be much higher than expected [27].

The carbon emission savings rate of ten different census regions in the four central regions in the United States reaches enormous dimensions, as seen in Fig. 1 on the left above. The Mountain South sub-region is the region that will contribute the most in this regard. Due to the continental climate conditions prevailing in this region, the rate of use of fossil resources is seen to be high in the sky [28]. The distribution across the United States by annual income is shown in Fig. 1 on the right above. According to this Figure, the lower-income group has been seen as the segment expected to contribute the most to carbon savings. As in this Figure, the middle-income and upper-income groups, defined as the 5th and 7th, are close to each other and can be matched with the standard daily living conditions. For this reason, changing lifestyles is vital in terms of seeing that consumption becomes regular after a while. The South Atlantic census region was selected for the analysis for this article due to its high propane usage (Fig. 2).

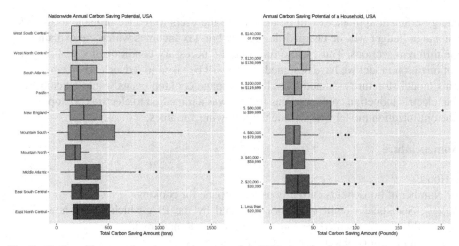

Fig. 1. Nationwide annual carbon saving potential, USA (on the left), annual carbon saving potential of a household by annual income, USA (on the right).

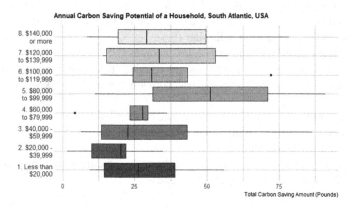

Fig. 2. Annual carbon saving potential of a household, South Atlantic, USA.

4.2 Energy Conversion

Energy conversion has become a parameter phenomenon, with energy efficiency being taken as a resource in future energy scenarios. Starting from the approach that today's most crucial issue is a carbon tax, we thought to place the main objective of transforming the cooktop ovens used in the kitchens in the context of the carbon tax. Specific to the RECS data set we used, we saw that the carbon emissions were reduced by 1.2% in the results of the first applications through Excel solver. Based on these trials, we think that evaluating the general strategy of the model within the framework of the energy efficiency transformation in the kitchens of the resources that will come from the carbon tax will have unexpected results at the first stage.

Firstly, the main objective of implementing this model is to encourage users who cook with propane in kitchens to use energy-efficient electric induction cooktop ovens and help reduce carbon emissions in this direction. As a result of various initiatives, energy-efficient appliances will be donated to the users to reduce propane usage. The campaign budget will be charged through carbon tax incomes which are explained in the different sections. Thus, while fulfilling the necessary conditions for realizing the optimization model we have designed, a step will be taken to reduce carbon emissions. This model structure assumed that all the propane cooktop ovens would be converted into electric induction cooktop ovens if there was a transition to electric cooktop ovens. The optimization model will include the following variables.

Nomenclature

i: Houses
N: Number of households currently using propane for cooking
N_{weight}: Statistical weight multiplier provided by the EIA which shows how many households are represented by a specific survey result nationwide
n_i^{stove}: Number of cooktop ovens in a household
C_{pe}: Cost of an electric-induction cooktop oven (USD)
C_{tax}: Carbon tax amount based on carbon saving potential (USD/pound of CO_2)
P_i^{tax}: Carbon tax payment of a specific household (USD)
MPC_i: Carbon tax multiplier for a household, based on the income level (%)
MIR_i: Household incentive ratio multiplier (%)
B_C: Total budget created with the carbon tax incomes (USD)
e_p: Carbon emission caused using propane for cooking (pounds per thousand BTU)
e_e: Carbon emission caused using electric cooktop ovens for cooking (pounds per thousand BTU)
u_i^p: The average monthly amount of propane used by the ith household for cooking (BTU)
u_i^e: The average monthly electricity usage for cooking in case the ith household switches to electricity induction (BTU)
M_i^{trn}: the amount of CO_2 emission savings achieved by the transition from propane to electricity induction of the ith household (Pounds).

4.3 Optimization Objective

The main aim of the optimization model is to reduce carbon emissions by developing a policy that will encourage citizens to transform their energy usage behaviors in kitchens. Because while electric cooktop ovens provide 74% energy efficiency, induction cooktop ovens provide 84% energy efficiency. In addition, the efficiency rate of propane appliances is around 40% [29]. Therefore, if most users use propane switches to the use of electric induction cooktop ovens, the energy savings and carbon emission reduction will be achieved, and it will show the success of the optimization model.

The data set for the optimization model we are trying to implement in this article is based on the result report of the residential energy consumption survey conducted by the United States Energy Information and Administration Center (EIA) to measure the residential consumption survey (RECS) data of 2015. The report was published in 2017 and reviewed in 2018. The primary sampling selection began with the geographical regions of the United States first, followed by subsections allowing for geographical control of the sample allocated to the Census Departments. Precise constraints for estimates of average energy consumption per household were then applied at the national, Census District, and Census Division levels to improve sample allocation for geographic areas. The 2015 RECS design has 19 geographic areas [30]. Also, it is necessary to define general information about the electric induction cooktop oven product; a 30-in. cooker with thermo-mate IHTB774C code will be used in this modeling. The main features of this product are as follows [31]:

Brand name	Thermo-mate
Efficiency	High efficiency/84%
Part number	IHTB774C
Max energy output	7200 W
Voltage	240 V (AC)
Wattage	7200 W

The RECS data set shows the limit of application for this model; therefore, the cooktop oven donation will not be based on households' expectations but on their income scales. The ratio between income and propane consumption will show the taken position on the transformation order line. In this context, scale units should be transformed from kwh to BTU to find the energy efficiency according to selected model induction cooktop ovens. Therefore, the scale will be shown gr/m^3. Also, scale units will be transformed again from pounds to BTU [32].

Due to the different efficiency ratios among propane and electricity, the energy amount required for cooking u_i^e should be calculated with the following formula.

$$u_i^e = \frac{u_i^p \times 0.40}{0.84}$$

In the formula, e_p represents carbon emission caused by using propane for cooking. u_i^p the average monthly amount of propane used by a household for cooking. e_e Carbon emission caused by using propane for cooking, and u_i^e is the average monthly electricity usage for cooking in case the household switches to electricity induction.

4.4   The Optimization Model

The main objective function of the optimization model below shows the equation among the parameters related to the model.

$$Max\, z = \sum_i^N M_i^{trn} \times N_{weight}$$

M_i^{trn} represents the amount of CO_2 emission savings achieved by the transition from propane to electricity induction in this formula. N_{weight} is the EIA's statistical weight multiplier, which shows how many households are represented by a specific survey result nation-wide. According to the model, the aim is to maximize the carbon emission reductions, where N shows the total number of households currently using the propane for cooking at their kitchens.

$$M_i^{trn} = \left(e_p \times u_i^p\right) - \left(e_e \times u_i^e\right)$$

In this formula, e_p represents the carbon emissions caused using propane (pounds per thousand BTU), and u_i^p shows the average monthly amount of propane used by the household for cooking (BTU) in this model. e_e are carbon emissions caused by using electric cooktop ovens for cooking (pounds per thousand BTU), and u_i^e is the average monthly electricity usage for cooking in case the household switches to electricity induction (BTU). The function, which was created to calculate the amount of CO_2 emission savings achieved by the transition from propane to electricity induction, is the part obtained by multiplying the carbon emission caused by the propane used for cooking with the monthly average propane used by the household for cooking when the carbon emission from the use of electric cookers is switched to electricity induction of the household, It is found by subtracting the result obtained from the multiplication of the monthly electricity usage that will be used for cooking.

The model's objective function will be limited by the budget of the cooktop ovens provided for the households. The sum of the carbon taxes P_i^{tax} are used as the budget of the program B_C. The formulation is presented as follows.

$$B_C = \sum_i^N P_i^{tax} \times N_{weight}$$

This budget will be gathered by setting a Carbon Tax Amount that will be charged from the households based on the carbon emission saving potentials (M_i^{trn}). The relationship between the carbon tax and carbon emission saving potential is given in the formula below.

$$P_i^{tax} = C_{tax} \times M_i^{trn} \times MPC_i$$

According to this formula, to find the P_i^{tax} which represents the carbon tax payment of a specific household in USD, the nationwide specified carbon tax C_{tax} (USD/pound of CO_2) is multiplied with carbon saving potential M_i^{trn} (Pounds), and MPC_i which is the carbon tax multiplier for a household, based on the income level (%). MPC_i multipliers are presented at the end of this section.

When it comes to the usage of this budget, the transformation costs are related to the budget as expressed in the following formula.

$$\sum_i^N C_{pe} \times n_i^{stove} \times MIR_i \times N_{weight} < B_C$$

The equation above represents that the budget collected by the carbon taxes will be used entirely for the transformation costs where C_{pe} is the cost of an electric-induction cooktop oven (USD) and taken as \$500. n_i^{stove} is the number of cooktop ovens used in a household, and MIR_i is the household incentive ratio multiplier (%) that shows what percentage of an electric-induction cooktop oven will be compensated by the campaign budget. The MIR_i is adopted based on the income levels of the households as shown in the table below.

Table 1. Adopted tax and incentive multipliers for different income levels (%)

Annual gross household income for the last year (MONEYPY)	Adopted tax multiplier	Adopted incentive multiplier
Less than \$20,000	0%	100%
\$20,000–\$39,999	0%	100%
\$40,000–\$59,999	0%	100%
\$60,000 to \$79,999	40%	75%
\$80,000 to \$99,999	80%	75%
\$100,000 to \$119,999	100%	75%
\$120,000 to \$139,999	100%	50%
\$140,000 or more	100%	50%

The data in Table 1 were taken from the RECS data and were created following the classification in the dataset. However, it must be combined with the national income tax system in the United States. Table 2 shows this combination according to RECS data offers. The national income tax system in the United States is designed in a multidimensional structure. Therefore, tax percentages vary depending on the difference between income levels. First, eight different income groups were classified in the RECS data which is processed through this article. Although the people's annual income levels increase, the amount of energy used becomes constant or similar after a while, so RECS has followed such an approach in classification. Therefore, while constructing the modeling in this article, it was necessary to combine the classification used in the RECS data and the tax classification prepared by the United States Internal Revenue Service. The data in Table 1 were taken from the RECS data and were created following the classification in the dataset. However, it must be combined with the national income tax system in the United States. Table 2 shows this combination according to the RECS data offer.

According to Table 3, to define the primary variable of the model, the carbon tax applied in this model has been determined as approximately \$29.54 for households with middle and upper-level household incomes. In addition, the exact product model will be donated to each home, and the market price of this product is approximately \$500. The entire carbon tax that is likely to be collected in total is considered as the campaign budget. This amount is foreseen as 524 million dollars on an annual basis. The main aim is for each household to receive at least one cooktop oven grant.

Table 2. The combined tax policy scale between annual income in recs data and the national tax percentage scale [33].

Annual gross household income for the last year (MONEYPY)	Single annual income	Married filing separately annual income	Married filing separately annual income	Head of household annual income
Less than $20,000	$0–$9,875	$0–$19,750	$0–$9,875	$0–$14,100
$20,000–$39,999	$9,876–$40,126	$19,751–$80,250	$9,876–$40,125	$14,101–$53,700
$40,000–$59,999	$40,126–$85,525	$80,251–$171,050	$40,126–$85,525	$53,701–$85,500
$60,000 to $79,999	$40,126–$85,525	$80,251–$171,050	$40,126–$85,525	$53,701–$85,500
$80,000 to $99,999	$85,526–$163,300	$171,051–$326,600	$85,526–$163,300	$85,501–$163,300
$100,000 to $119,999	$85,526–$163,300	$171,051–$326,600	$85,526–$163,300	$85,501–$163,300
$120,000 to $139,999	$85,526–$163,300	$171,051–$326,600	$85,526–$163,300	$85,501–$163,300
$140,000 or more	$85,526–$163,300	$171,051–$326,600	$85,526–$163,300	$85,501–$163,300
Annual gross household income for the last year (MONEYPY)	Adopted tax multiplier	Adopted incentive multiplier	Annual income tax percentage annual income	Carbon tax policy percentage
Less than $20,000	0%	100%	10%	0%
$20,000–$39,999	0%	100%	12%	0%
$40,000–$59,999	0%	100%	22%	0%
$60,000 to $79,999	0%	75%	22%	8%
$80,000 to $99,999	80%	75%	24%	8%
$100,000 to $119,999	100%	75%	24%	8%
$120,000 to $139,999	100%	50%	24%	8%
$140,000 or more	100%	50%	24%	8%

Table 3. The details related to the optimization model in Table 3 above

Objective	Maximize CO_2 emission reduction	14.873	Tons of CO_2 (annually)	
Fixed amount	Cost of one oven	$500	USD	
Variable	Carbon tax amount	$30	USD/Pound	
Constraint	Total campaign budget	$523.814.715	<	$523.814.715

4.5 Model Analysis

Since the purpose of the Zero Carbon Energy Transition optimization modeling is to maximize to reduce the existing carbon emissions in the kitchens, an analysis will be presented in the context of the results obtained because of processing the data from all states of the South Atlantic region of the United States in the RECS dataset and processing it in an excel solver. In this context, the first findings regarding the results obtained with the application can be seen as follows.

Firstly, the correlation was established between the income levels of the household and the carbon tax distribution. This correlation aims to see how the changes in living standards depending on the increase or decrease in the household's income level change the energy consumption trends in the South Atlantic region. In addition, since the changes in the regions and sizes of the households are parallel with the change of trends, it is necessary to understand how consumption takes shape in which areas.

In the South Atlantic Region of the United States, it has been seen that the income groups that emit the most carbon, depending on the change in living conditions, are the American middle and the middle-high classes. Accordingly, the 7th group, which is shown in Fig. 3 on the left below, whose income level is 120–140K the US $, became the group that will pay the most tax payment to implement the carbon tax. Later, the 5th and 8th groups follow the 7th. It is necessary to exclude all groups whose income is below $40k from the carbon tax. Since these groups have difficulty providing standard living conditions according to the United States tax system which is shown in Table 2, it is unlikely that they will be able to lift such a tax burden. For this reason, priority will be given to benefit from the grant program to reduce the carbon emissions created by other groups thanks to the tax. The carbon tax distribution in the South Atlantic by annual household income level is shown in Fig. 3 on the right below.

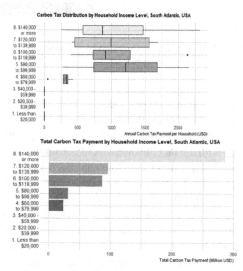

Fig. 3. Annual carbon saving potential of a household, South Atlantic, USA (on the left), total carbon tax payment by household income level, South Atlantic, USA (on the right).

However, as can be seen in Fig. 3 on the right above, it is understood that the 8th group is the group that pays the most tax when the carbon tax application is introduced. This is because the households' consumption in that group in the kitchens is higher than that of the 7th group. In addition, another critical point here is that, as seen in Fig. 3 on the left above, the 5th group is the group that will make the most significant contribution to reducing carbon emissions because this income group is more numerous than other income groups. All this network of mutual relations reveals that to reduce carbon emissions at the maximum level, a transition should be provided for an income group and all income groups with an optimal approach according to consumer trends. Again, as shown in Fig. 3 on the right above, the carbon tax collected from the 8th group is higher than the total amount of carbon tax collected from the 7th, 6th, 5th, and 4th groups. This situation gives us both a prediction about the living standards of that group and information about the size of the consumption amounts.

As a result, the *3rd Figure* provided the understanding of the most significant tax resource group of the carbon tax application approach, which is the subject of this modeling analysis. Therefore, both the entire group with an income below $40K and some of the other groups' carbon conversion expenses will be covered by the amount provided from the 8th group, which is seen as the leading tax source.

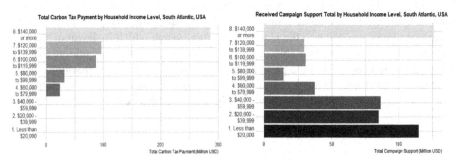

Fig. 4. Total carbon tax payment (on the left), received campaign support total by household income level, South Atlantic, USA (on the right).

Considering the total campaign support received by household income level, the lowest group was the group that received the most support. This should not be surprising, as this is already an expected situation. In addition, the 5th group receiving the least support here is associated with the fact that household consumption is close to the carbon tax limits. It must remind that a new cooktop oven has been allocated to each household in the distribution made from the total carbon tax revenues. The direct proportional approach between the decrease in the amount of income and the decrease in energy consumption is also seen in Fig. 4 on the right above. Therefore, starting from the idea that living standards are the main variable determining consumption curves, the urgent transformation of all classes below the middle class will play an essential role in reducing carbon emissions. Figure 4 on the right above also predicts the income distribution of the South Atlantic region when examined from an economic perspective. The lowest income group is also the group that needs the most conversions.

4.6   Regression Analysis in R

The second application of the model was performed in the R program in data from all other states of the United States nationwide. Seeing the distribution of all states in the RECS dataset may be necessary to understand applying the Carbon tax in the South Atlantic. It is, therefore, needed to present the definition of the five main variables.

NHSLDMEM: Number of household members
TOTSQFT_EN: Total square footage (used for publication)
URBANTYPE: Census 2010 Urban Type
INCOMEAVERAGE: Annual gross household income for the last year (Numeric Average)
COOKTOPNEED: Number of cooktop ovens

Table 4. Residual coefficients of RECS data nationwide.

```
Residuals:
    Min      1Q   Median      3Q      Max
-44.665 -17.350   -5.863  10.075  146.769

Coefficients:
                        Estimate Std. Error t value Pr(>|t|)
(Intercept)            3.202e+01  5.895e+00   5.433 1.17e-07 ***
NHSLDMEM               3.288e+00  1.093e+00   3.008  0.00286 **
ATHOME                 1.687e+00  8.023e-01   2.102  0.03639 *
TOTSQFT_EN             6.757e-04  1.211e-03   0.558  0.57738
URBANTYPEUrban Area   -3.474e+00  3.617e+00  -0.960  0.33759
URBANTYPEUrban Cluster -4.118e+00 6.112e+00  -0.674  0.50098
INCOMEAVERAGE         -2.040e-06  3.929e-05  -0.052  0.95862
COOKTOPNEED           -9.920e+00  3.841e+00  -2.583  0.01028 *
---
Signif. codes:  0 '***' 0.001 '**' 0.01 '*' 0.05 '.' 0.1 ' ' 1

Residual standard error: 25.79 on 294 degrees of freedom
Multiple R-squared:  0.06828,   Adjusted R-squared:  0.04609
F-statistic: 3.078 on 7 and 294 DF,  p-value: 0.00382
```

The number of household members in energy consumption could be seen as effective, as seen in Table 4. Suppose it is accepted that the sensitivity value increases as it approaches "0.001" with a value like "0.00286". In that case, it can be said that the change in the number of households is a determinant in the consumption function.

The increase in the number of households and consumption is directly proportional, but it is not algorithmic. For this reason, the accumulation in consumption occurred in groups with 1 to 3 households. As seen in Fig. 5 on the left above, the aggregation was the highest in households with 2 people. Since the number of cooktop ovens in the house has not changed much despite the increase in the number of households, it can be said that it is openly expected that consumption has increased slightly. The directly proportional relationship between the increase in the number of households and carbon emissions is seen in the same way in the South Atlantic and throughout the United States.

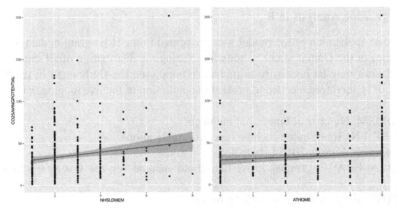

Fig. 5. Carbon saving potential by number of household (on the left), and by home (on the right), the USA

For this reason, as seen in Fig. 5 on the right, the savings potential of carbon emissions is highest in places where the number of households is high. The similarity between the households' potential carbon emission savings rates with less than 5 (number of people) is also related to the number of cooktop ovens in the inn.

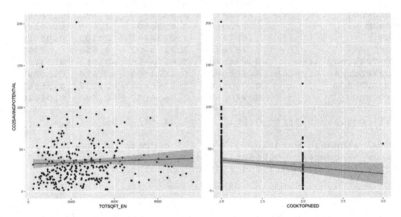

Fig. 6. Annual carbon saving potential by total square footage (on the left), and by cooktop oven need (on the right), USA

The consumption amount, which varies depending on the number of kitchens in the house, not the size of the houses, has accumulated below 4000 SQF, as seen in Fig. 6 on the left. Because this data, calculated on the average size of households across the United States, revealed that consumption does not depend on the size of the house.

Finally, Fig. 6 shows that each household needs at least a cooktop oven to prepare cooks. The carbon saving rate of those who use more than one cooktop oven may not be seen much differently because households cannot use those ovens every time. No matter which household has more than one cooktop oven, it should not be expected to significantly contribute to the savings amount since it is not possible to use both continuously.

5 Conclusion

This article presents an optimization model that aims to reduce carbon emissions at the maximum level by financing the conversion of cooktop ovens in households' kitchens in the South Atlantic region using the fund created by the carbon tax policy. The expenditure of all the income collected through the carbon tax for the transition from propane to electric induction in kitchens will have significant effects on global warming, climate change, and energy efficiency. It has been observed that there is a 1.2% carbon emission saving. Potential even with the optimization made only for the South Atlantic region. It has been revealed that the efficiency loss arising from the use of propane will be prevented within the scope of this transformation. This optimization model can be considered a pilot project.

The RECS data is used in the model provided sufficient parameters for the model to work due to its both quantitative and qualitative variable properties. In this respect, relevant optimization models can be developed using both R and Excel. The primary formulation of the model can be realized by giving at least one cooktop oven to all regional households, regardless of household type. The formulation of carbon tax implementation should consider the annual income and energy consumption volumes to finance conversions with equal rights. With this approach, in line and the recommendations of the United States Department of Internal Revenue, a tax-free model has been developed for income groups below $40k, and it has been revealed that culinary transformations can be realized for all members of this lower-income groups within the scope of its implementation. Moreover, since the data is classified over different income groups, it is suitable for processing with statistical programs.

It is recommended that we realize the same application for the poorer towns of Turkey by collecting the data. However, more data than 300 used in this study will cause limits in using Excel, which would be totally converted to Gurobi functions of R software. Explicit carbon-pricing instruments can raise revenue efficiently because they help overcome a key market failure: the climate externality. The revenue can be used to foster growth in an equitable way by returning the revenue as household rebates, supporting poorer sections of the population, managing transitional changes, investing in low-carbon infrastructure, and fostering technological change. Ensuring revenue neutrality via transfers and reductions in other taxes could be a policy option. Policy decisions will need to take into account the country's objectives duly and specific circumstances, while keeping in mind the development objectives and commitments agreed in relation to the Paris Agreement objectives [34].

Appendix

See Table 5.

Table 5. Federal income tax brackets for 2021 (filling deadline: April 15, 2022).

	Single	Married filing jointly	Married filing separately	Head of household
10%	$0–$9,950	$0–$19,900	$0–$9,950	$0–$14,200
12%	$9,951–$40,525	$19,901–$81,050	$9,951–$40,525	$14,201–$54,200
22%	$40,526–$86,375	$81,051–$172,750	$40,526–$86,375	$54,201–$86,350
24%	$86,376–$164,925	$172,751–$329,850	$86,376–$164,925	$86,351–$164,900
32%	$164,926–$209,425	$329,851–$418,850	$164,926–$209,425	$164,901–$209,400
35%	$209,426–$523,600	$418,851–$628,300	$209,426–$314,150	$209,401–$523,600
37%	$523,601+	$628,301+	$314,151+	$523,601+

References

1. Barron, A.R., Fawcett, A.A., Hafstead, M.A., McFarland, J.R., Morris, A.C.: Policy insights from the EMF 32 study on U.S. carbon tax scenarios. Climate Change Econ. **9**(1), 1840003 (47) (2018)
2. Metcalf, G.E.: The distributional impacts of U.S. energy policy. Energy Policy **129**, 926–929 (2019)
3. King, M., Tarbush, B., Teytelboym, A.: Targeted carbon tax reforms. Eur. Econ. Rev. **119**, 526–547 (2019)
4. Goulder, L.H., Hafstaed, M.A., Kim, G., Long, X.: Impacts of a carbon tax across US household income groups: what are the equity-efficiency trade-offs? J. Public Econ. **175**, 44–64 (2019)
5. Bhandari, V., Giacomoni, A.M., Wollenberg, B.F., Wilson, E.J.: Interacting policies in power systems: renewable subsidies and a carbon tax. Electr. J. **30**(6), 80–84 (2017)
6. Ghaith, A.F., Epplin, F.M.: Consequences of a carbon tax on household electricity use and cost, carbon emissions, and economics of household solar and wind. Energy Econ. **67**, 159–168 (2017)
7. Geroe, S.: Addressing climate change through a low-cost, high-impact carbon tax. J. Environ. Dev. **28**(1), 3–27 (2019)
8. Hafstead, M.A., Chen, Y.: Using a carbon tax to meet US international climate pledges. Resources for the Future (RFF) (2016)
9. Ramseur, J.L., Leggett, J.A.: Attaching a price to greenhouse gas emissions with a carbon tax or emissions fee: considerations and potential impacts. Homeland Secur. Digit. Libr. (2019). https://www.hsdl.org/?abstract&did=823346. Accessed 24 June 2021
10. Metcalf, G.E.: On the economics of a carbon tax for the United States. Brookings Papers on Economic Activity, no. Spring, pp. 405–458 (2019)
11. Aldy, J.E.: Carbon tax review and updating: institutionalizing an act-learn-act approach to U.S. climate policy. Rev. Environ. Econ. Policy (Winter) (2020)

12. Lucas, G.M., Jr.: Behavioral public choice ABD the carbon tax. Utah Law Rev. **115** (2017)
13. Kotchen, M.J., Turk, Z.M., Leiserowitz, A.A.: Public willingness to pay for a US carbon tax and preferences for spending the revenue. Environ. Res. Lett. 094012(12) (2017)
14. McCauley, D., Heffron, R.: Just transition: integrating climate, energy and environmental justice. Energy Policy **119**, 1–7 (2018)
15. Boussemart, J.-P., Leleu, H., Shen, Z.: Worldwide carbon shadow prices during 1990–2011. Energy Policy **109**, 288–296 (2017)
16. Stavins, R.: The future of U.S: carbon pricing policy. NBER Working Paper Series, no. Working Paper 25912 (2019). http://www.nber.org/papers/w25912
17. Horowitz, J., Cronin, J.-A., Hawkins, H., Konda, L., Yuskavage, A.: Methodology for analyzing a carbon tax. The Department of the Treasury, no. Working Paper 115 (2017)
18. Shapiro, A.F., Metcalf, G.E.: The macroeconomic effects of a carbon tax to meet the U.S Paris agreement: the role of firm creation and technology adoption. Resources for the Future, no. Working Paper 21–24 (2021)
19. Hafstead, M.A., Williams, R.C., III.: Designing and evaluating a US carbon tax adjustment and mechanism. Resources for then Future, no. Working Paper 20–24 (2020)
20. Metcalf, G.E.: An Emissions Assurance Mechanism: Adding Environmental Certainty to a Carbon Tax. Resources for the Future, Washington DC (2018)
21. Cooper, M.: Governing the global climate commons: the political economy of state and local action the U.S flip-flop on the Paris Agreement. Energy Policy **118**, 440–454 (2018)
22. McKibbin, W.J., Morris, A.C., Wilcoxen, P.J., Liu, W.: The role of border carbon adjustments in a U.S carbon tax. Climate Change Econ. **9**(1), 1840011 (41 pages) (2018)
23. Metcalf, G.E., Stock, J.H.: Measuring the macroeconomic impact of carbon taxes, Harvard Environmental Economics, no. Discussion Paper, pp. 20–86 (2020)
24. Haites, E.: Carbon Taxes and Greenhouse gas emissions trading systems: what have we learned? Climate Policy **18**(8), 955–966 (2018)
25. Wang, Q., Mu, R., Yuan, X., Ma, C.: Research on energy demand forecast with LEAP model based on scenario analysis - a case study of Shandong province. In: 2010 Asia-Pacific Power and Energy Engineering Conference, Chengdu, China (2010)
26. Wan, Wan, J.: US-China Trade war: speculative saving perspective. Chinese Econ. **54**(2), 107–123 (2021)
27. Each Country's Share of CO2 Emissions: Union of Concerned Scientists, 12 August 2020. https://www.ucsusa.org/resources/each-countrys-share-co2-emissions. Accessed 24 June 2021
28. U.S. Energy Information Administration - EIA - Independent Statistics and Analysis: Fossil fuels account for the largest share of U.S. energy production and consumption - Today in Energy - U.S. Energy Information Administration (EIA), 14 September 2020. https://www.eia.gov/todayinenergy/detail.php?id=45096. Accessed 24 June 2021
29. Matthew, L.: Which is More Energy Efficient - Gas, Electric, or Induction? LeafScore, 15 March 2021. https://www.leafscore.com/eco-friendly-kitchen-products/which-is-more-energy-efficient-gas-electric-or-induction/. Accessed 24 June 2021
30. U.S. Energy Information Administration - EIA - Independent Statistics and Analysis: Residential Energy Consumption Survey (RECS) - Energy Information Administration, 31 July 2018. https://www.eia.gov/consumption/residential/. Accessed 24 June 2021
31. Electric Cooktop Thermomate: Electric Cooktop, thermomate 30 Inch Built-in Induction Stove Top (2020). https://www.amazon.com/Electric-thermomate-Induction-Smoothtop-Cer tificated/dp/B08CXMQMM6. Accessed 24 June 2021
32. Energy Units: American Physical Society. https://www.aps.org/policy/reports/popa-reports/energy/units.cfm. Accessed 24 June 2021
33. Home: Internal Revenue Service: Internal Revenue Service | An official website of the United States government, 30 April 2021. https://www.irs.gov/. Accessed 24 June 2021

34. The World Bank: Report of the High-Level Commission on Carbon Prices (2017)
35. Federal Income Tax Brackets for Tax Years 2019 and 2020: SmartAsset, 30 April 2021. https://smartasset.com/taxes/current-federal-income-tax-brackets. Accessed 20 June 2021

Barriers and Challenges of Knowledge Management in a Gas Company

Piotr Domagała$^{(\boxtimes)}$ (ID)

Wroclaw University of Economics and Business, Komandorska, 118/120, 53-345 Wrocław, Poland
piotr.domagala@ue.wroc.pl

Abstract. Gas is a greener alternative to coal. It is used to heat houses as well as fuel for cars. The gas industry market is dominated by large companies, operating across countries and continents. It is constantly growing. Managing such a large companies is extremely complex due to the scale of operations, rapid development of technologies and a large number of employees who work remotely and are geographically dispersed. In the context of the increasingly common treatment of knowledge as one of an organization's key resources and growing number of data generated by business, Knowledge Management has received considerable attention in the energy sector, including gas sector because of its impact on their performance. It is not an easy task because of complexity of whole process.

The author points out the potential barriers to the development of Knowledge Management in a gas company according to carried research.

Keywords: Knowledge management · Gas company · Competitive advantage

1 Introduction

Caring about the environment is one of the foundations of sustainable development. Nowadays, our civilization struggles, among others, with the problem of air pollution. Gas replaces coal with less harmful gas as a heating fuel. Gas is also a very popular solution used in the automotive industry. Whole industry develops extremely fast as a consequence.

The gas market is dominated by large enterprises that operate across entire countries and continents. Managing such a large companies is very complex due to the scale of operations, rapid development of technologies and a large number of employees who work remotely and are geographically dispersed.

There is a huge challenge in terms of the aging workforce as a large number of employees near retirement will soon leave the industry, thus begetting a profound knowledge loss. Layoffs will also occur when businesses are not performing well [1].

The existing literature does not provide detailed insights into the challenges and strategies associated with knowledge retention. There is also a lack of research on the global perspective of knowledge retention activities of retiring employees.

© IFIP International Federation for Information Processing 2022
Published by Springer Nature Switzerland AG 2022
E. Mercier-Laurent and G. Kayakutlu (Eds.): AI4KMES 2021, IFIP AICT 637, pp. 63–74, 2022.
https://doi.org/10.1007/978-3-030-96592-1_5

The author presents impact of an ageing workforce on knowledge retention, barriers that significantly hinder the knowledge management process in a gas company and gives examples of solutions to improve knowledge retention and, consequently, the entire knowledge management process.

In the first chapter, the author focuses on the essence of knowledge management success and its growing importance for business In the second section elements of Knowledge Management process are identified. In the last section, the author indicates the problems and needs in KM area in the gas company according to the research.

2 Role of Knowledge Management in Business

2.1 Knowledge as a Factor of Successful Development

Knowledge is the foundation of economic growth today. Producing goods and services with high added value is the key to improving economic performance and enhancing international competitiveness. Increasing efforts in this direction has become a fundamental challenge for enterprises [2].

According to the theory of knowledge-based organizations, knowledge by itself does not create a competitive advantage, and it is achieved through its application and integration with the company's business processes [3].

Knowledge Management (KM) is, in part, an attempt of the best possible use of knowledge, which is available in organization, creation of new knowledge and growth of knowledge understands [4].

Knowledge should be regarded as one of the strategic resources of the company, on a par with human resources, financial resources and material resources (equipment) and needs to be managed in appropriate way. It leads to Knowledge Management evolution and dissemination of intelligent organization concept.

2.2 Knowledge Management Evolution

Dynamic changes in the environment force changes in KM strategy and philosophy. KM is no longer an extra work to do. It is helpful to do our work in a real time.

Changes in the approach to management and the ever-growing role of knowledge in organizations go hand in hand with changes in the approach to Knowledge Management.

The table below shows changes in the approach to Knowledge Management over the years (Table 1).

At the beginning KM was targeted on knowledge collecting and the main tool were Knowledge Management Systems. Whole process of knowledge management was treated as an extra work.

With the passage of time and the development of web 2.0 technologies and social media, the approach has changed. It started to focus on the social importance of knowledge sharing.

Nowadays, from technical point of view, collecting and sharing of knowledge is not a big deal. People generate huge amounts of data every day. The problem is how to choose the most valuable information to support our work, make it useful and let company achieve competitive advantage.

Table 1. Knowledge management evolution.

KM 1.0	KM 2.0	KM 3.0
Collecting techno-centric approach	**Sharing** people-centric approach	**Using** productivity-centric approach
— traditional organizational — content/community — approaches — command and control — "KM is an extra work" — Focused on collecting knowledge — „before it walked at the door"	— enriched with social media and ecosystem wide co-creation "KM is part of my work" — social "KM is a part of my work" — Focused on sharing knowledge by using web-enabled and Social Media tools	— KM augmented with AI — practical & individual „KM is helping my to do my work" — Focused on using existing knowledge to help people get their job done

2.3 Intelligent Organization Concept

The author mentioned that knowledge treated as one of strategic resources needs to be managed in appropriate way what leads to dissemination of intelligent organization concept.

Intelligent organization is assumed that it is an organization bases its philosophy on Knowledge Management. This term has become so popular due to the rapid development of ICT technologies, dynamically changing economic environment and increased market competitiveness. We can consider organization as intelligent when it is a learning organization with the ability to create, acquire knowledge, organize, share it and use it to increase efficiency and competitiveness on the global market [4] (Fig. 1).

As the author shows at the picture above, growing number of usable data, knowledge treated as a valuable resource for business and develop of Knowledge Management are engine of modern, intelligent organization.

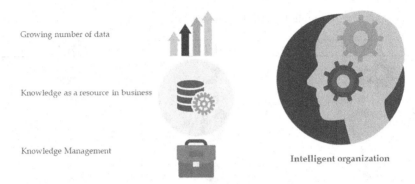

Growing number of data

Knowledge as a resource in business

Knowledge Management

Intelligent organization

Fig. 1. Intelligent organization concept.

3 Knowledge Management in a Gas Company

3.1 Aging Workforce and Knowledge Retention in a Gas Company

Knowledge retention has become an important and inevitable activity in organizations these days due to demographic changes and the graying of employees. There is a serious threat to the organizations for knowledge loss when employees leave [6].

As Fig. 2 depicts, oil, gas and mining companies reported approximately one-quarter (27%) workforce in the industry is age 55 of older, similar to many other industries. About one-fifth (17%) of HR professionals in oil, gas and mining companies said they were not aware that the proportion of older workers was increasing and that older workers were projected to make up approximately 26% of the labor force by the year 2022, compared with 21% in 2012 and 14% in 2002 [7]. Losing these workers means organizations lose the much-needed knowledge which is basis for their competitive advantage. The knowledge of these employees is of key importance, as it may lead to a decay of organizational memory when these employees leave, which in turn may reduce the company's ability to identify and use past knowledge for competitive advantage.

If we take into account factors such as the geographical dispersion of workers, their large number and the duties they perform, we can certainly conclude that they generate a significant amount of information and are in possession of knowledge of a very different nature. It is required to properly analyze the skills and capabilities of these employees, as they might be working on different job assignments during their careers and have a variety of expertise in different areas within the organization [8]. The comparison and analysis of known models of knowledge management processes allowed to propose four main problem areas related to the circulation of knowledge in the organization. These are:

- Identifying and locating knowledge,
- Knowledge retention,
- Knowledge transfer,
- Knowledge utilization.

The author describe them in the following sections of the chapter.

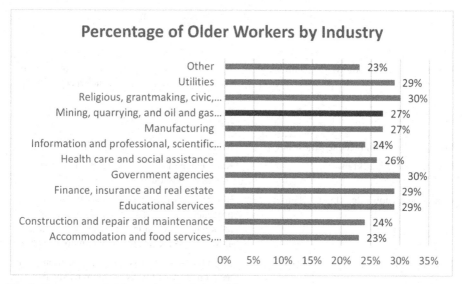

Fig. 2. Percentage of older workers by industry **Source:** Preparing for an Aging Workforce: Oil, Gas and Mining Industry Report (SHRM 2015)

3.2 Identification of Critical Knowledge

Critical knowledge, is the unique types of knowledge and skills possessed by the employees whether it is knowledge related to management, technical areas or relationships [9]. Alavi and Leidner describe a knowledge taxonomy providing a possible context for considering the different knowledge types that the departing employees might possess [10]. The Table 2 shows a breakdown of knowledge by object.

We can also divide knowledge according to its importance to the organization:

- technological key knowledge (knowledge of the main processes that generate knowledge for the customer),
- coordination key knowledge (knowledge of the resources, tools and solutions related to the management of knowledge in the organization),
- auxiliary knowledge (knowledge covering the supporting processes in the company),
- market knowledge (knowledge about the market segments in which the organization operates, customer needs and how to meet them).

Identification is a key activity for successful knowledge management processes. Many researchers point out the dangers of missing or inadequately identifying the knowledge of an organization and its environment. Hence, it is extremely important to organize it according to clearly defined criteria.

Identifying knowledge aims to identify not only the current, but also the target, postulated and ideal state of knowledge in a company. The difference between the current state and the desired state is the knowledge gap.

Table 2. Types of knowledge by object.

Type of knowledge	Description
Declarative/explicit knowledge (know about)	A codified, tangible/physical description of the knowledge in question to which a company has ownership rights Factual knowledge, very close to the concept of information
Procedural/tacit knowledge (know how)	Knowledge relates to skills, ability to produce
Causal knowledge (know why)	Knowledge of cause and effect relationships and their consequences
Conditional knowledge (know when)	Refers to knowing when to use declarative and procedural knowledge. It allows to allocate resources when using strategies. This in turn allows the strategies to become more effective
Relational knowledge (know with)	Knowledge of other people, organizations who have knowledge that is useful to us and can help us to achieve our goals

3.3 Knowledge Retention

Within the process of knowledge retention/gathering we can distinguish between activities aimed at: capturing, processing, storing, securing and protecting the organization's knowledge.

Identified and localized knowledge should be extracted to serve the objectives of the organization and its members. Its processing should enable access to the organization's knowledge resources in an accessible and understandable form for all stakeholders. The storage of knowledge will serve to preserve it within the organization for repeated use in the future, and its preservation and protection are intended to ensure that the knowledge is used in a manner consistent with the organization's goals and policies.

We consider the problem of knowledge accumulation in an organization from two aspects - the technological aspect (advanced ICT technology in the form of database systems, expert systems, knowledge bases/portals, etc.) and the human aspect (knowledge accumulation in people).

The former is accentuated in organizations focused on explicit knowledge, aiming to codify and document employee knowledge. In the second case, the emphasis is on the soft aspects, the absorption of knowledge and the learning of individuals (Fig. 3).

The figure above summaries the most significant differences in knowledge codification and personalization strategies.

Many experts wonder which knowledge management tools to use in the context of the strategies outlined above. The results of a study by J. Román Velázquez can be used for this purpose, the results of which are presented in the figure below [11] (Fig. 4).

The diagram shows how different tools can be used due to the company's preferred Knowledge Management strategy.

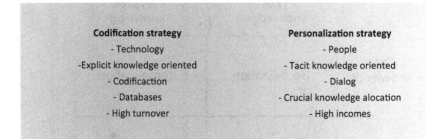

Codification strategy	Personalization strategy
- Technology	- People
-Explicit knowledge oriented	- Tacit knowledge oriented
- Codificaction	- Dialog
- Databases	- Crucial knowledge alocation
- High turnover	- High incomes

Fig. 3. Codification and personalization strategy.

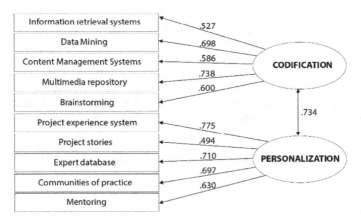

Fig. 4. Knowledge management strategies and the tools they use.

3.4 Knowledge Transfer

The term knowledge transfer refers to activities, tools and conditions that enable the flow and sharing of knowledge within an organization.

The academic literature shows a dualism between the concepts of knowledge flow in organizations, which is often considered independently of the concept of knowledge sharing. These concepts are often equated with each other, but there are scientific indications that these concepts have different meanings [12]. Researchers point out that knowledge transfer is characterized by, among other things, an unambiguous purpose, unidirectional transfer and concentration and focus. In terms of the above characteristics, the scientific world agrees.

It is worth noting at this point the important role of the transformation model and the mutual transformation of explicit and tacit knowledge (Fig. 5).

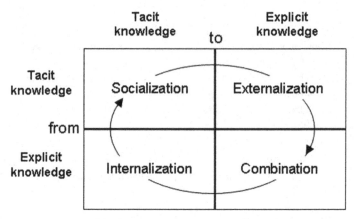

Fig. 5. Knowledge transformation.

Nonaka assumed that knowledge growth is possible through the occurring relations of tacit and explicit knowledge [13]. Based on this assumption, he identified 4 types of transformation of knowledge types, which are shown in the diagram above.

Table 3 presents these relations/transformations in details.

Table 3. Relations in knowledge transformation.

Transformation	Description
Socialization	Tacit → Tacit The creation and transfer of tacit knowledge during face-to-face interactions between individuals It is reflected in staff development methods such as on-the-job training or manager shadowing
Combination	Explicit → Explicit Creating new knowledge by reconfiguring existing explicit knowledge resources: sorting, adding, categorizing, giving new meanings It is reflected in the form of artefacts: documentation, databases and data warehouses, publications, internal manuals, instructions
Externalization	Tacit → Explicit Verbalizing tacit knowledge held, describing it using language, metaphors, models, action schemes, knowledge maps It allows knowledge to be extracted from the individual while preserving it in a form that can be used later
Internalization	Explicit → Tacit Internalization through the process of learning enriches the individual knowledge of employees with knowledge assimilated from explicit knowledge carriers This is equivalent to the concept of "learning by doing"

The concept has found wide recognition in the Knowledge Management world. The model created has become the basis for designing many Knowledge Management solutions and tools.

3.5 Knowledge Utilization

Putting knowledge into practice is a much more difficult activity than other knowledge management processes. Repeated observations by researchers (J. Pfeiffer and R.I. Sutton) support the finding that people understand the nature of the problem, know what needs to be done to improve company performance, but do not do it. Moreover, research shows that the success of the implementation of these measures does not depend on the introduction of new, hitherto unknown methods, but precisely on the ability to make practical use of the knowledge that already exists in the organization [14].

Using knowledge in practice it is not an easy task. In the next part of the publication, the author points out the most common causes of Knowledge Management failures, barriers and challenges in gas company.

4 Knowledge Management Barriers in Gas Company

Knowledge management processes are ancillary, supportive to the main business processes in organizations. Their contribution to reducing costs, increasing quality, shortening the time of execution of critical processes should be measurable and perceivable. Otherwise, all knowledge management efforts in an organization lose their justification.

The processes of identifying and locating knowledge, collecting and transferring it, presented by the author, should focus as much as possible on using the knowledge that is the subject of these processes, so that it becomes possible to quickly and efficiently apply the accumulated knowledge in practice to solve problem situations.

According to the theory of knowledge-based organizations, knowledge by itself does not create competitive advantage, it is achieved through its application and integration into companies' business processes [3].

The author presents what are the most common reasons for the failure of the knowledge management process in a gas company.

At the beginning the author described four main areas of problems in Knowledge Management (identification, retention, transferring and knowledge utilization). Then, the author use available research findings [3] outlining common barriers to knowledge management in a gas industry company and then conducted in-depth interviews with employees of a company in the same industry in Poland.

Among those surveyed were employees from: executives (1), HR (2), ICT (3), sales (4), planning & investment (5), real estate (6), finance (7), law (8), Field staff (9). In this way, the barriers for the company under study are identified.

For the purposes of the study, barriers were grouped thematically in areas: managing knowledge processes, people, technology, processes/organization, environment, characteristics of knowledge. Then the author rejected barriers that were not selected by the respondents in the initial stage of the survey and conducted an in-depth interview on the others.

The table below presents the results – barriers identified by employees in each business area of the organization (Table 4).

Table 4. KM barriers according to business areas – results.

	1	2	3	4	5	6	7	8	9
O: Lack of an appropriate reward for acquiring new skills and knowledge			+	+					+
T: Lack of technical support of integrated technology to support KM tools requirements	+	+		+	+	+	+	+	+
O: Lack of formal authority on the part of the innovator	+	+	+		+			+	+
O: Lack of fit between innovation and organizational assumptions, existing occupational mindsets, beliefs, etc.	+	+			+		+	+	+
O: Lack of teams, work groups and communities of practice for collaboration around work-related issues and challenges		+	+	+		+			+
P: Lack of slack times and heavy workload	+	+		+	+		+		+
O: Poor targeting of required information in order togenerate needed knowledge in KM system		+	+			+		+	+
K: Staff's ambiguity about what the information/knowledge is exactly supposed to be used for			+					+	+
P: Lack of top management support (supporting new ideas, establishing sufficient technical infrastructure, encouraging and so on)		+	+	+	+	+	+	+	+
O: Poor coordination among functions caused by top-down leadership style		+	+	+	+	+	+	+	+
O: Inconsistencies between strategy, systems, policies and practices		+	+					+	
T: Staff's reluctance to use integrated IT systems and tools due to lack of familiarity and experience with systems and tools	+	+							+

Question 1: In which group of employees were the most barriers identified?
    Answer: Field Stuff, HR, IT & Law.
Question 2: What barriers were identified by employees most frequently?
    Answer: Lack of technical support of integrated technology to support KM tools requirements, Lack of top management support, Poor coordination among functions caused by top-down leadership style.
Question 3: What group of barriers were identified by employees most frequently?
    Answer: Processes/Organization (O).

Field workers (assemblers) are among the group of workers who indicated the most barriers related to knowledge management in their area of operation. The barriers they indicated relate mainly to organizational and technical issues. This may be related to the average lowest level of education of this group of respondents and low awareness of knowledge management needs. These employees are in possession of high technical knowledge. The low competence in the area of Knowledge Management by HR staff is worrying, as they make strategic decisions in the company. The high number of barriers indicated by the IT area, especially in the organizational aspect, is most probably caused by the large number of supported systems of a multiple nature (from strictly industrial systems such as SCADA systems, through SAP to office applications such as MS Office), which results in disorganized knowledge.

It is noteworthy that among the barriers most frequently mentioned by employees are those that relate to organizational issues. Both in terms of personnel and technical issues.

5 Conclusions

As the universally accepted and well-documented results of scientific research show, humanity, in order to avoid the most severe global warming, must radically reduce its greenhouse gas emissions by half within 10 years, and ultimately to zero by 2050.

The need for a rapid phase-out of fossil fuels in the energy sector is not only due to climate concerns, but also for economic reasons.

Peak sources to replace coal-fired power stations include gas-fired power stations. Modernization and expansion of transmission networks will be of key importance, including their adaptation to the increasing dispersion of electric energy sources. This means an increase in interest in gas on the market and therefore an increase in competitiveness. This means an increase in interest in gas on the market and therefore an increase in competitiveness. Companies wishing to build their market position and competitive advantage will compete in the area of the development of new technologies. Launching of new solutions is closely linked to the acquisition, possession and use of specialist knowledge. Knowledge has already become one of the key resources in organizations and its role is constantly growing. This process cannot take place without effective management.

As a study conducted in one of the leading companies in the gas sector in Poland shows, there is a lot to be done in this area.

The basis is understanding the essence of knowledge management and treating this process as one of the priorities. Many years of neglect in this area cause lack of awareness and knowledge of tools for effective acquisition, consolidation and transfer of knowledge. According to ageing workforce, action to improve knowledge management in the gas industry needs to be taken immediately, otherwise key knowledge will be irretrievably lost along with a large group of departing employees.

References

1. Aggestam, L., Durst, S.: Using IT-Supported knowledge repositories for succession planning in SMEs: how to deal with knowledge loss?. In: Handbook of Research on Human Resources Strategies for the New Millennial Workforce. IGI Global (2017)
2. Amidon, D., Skyrme, D.: The knowledge agenda. J. of Knowl. Manage. 1(1), 27–37 (1997)
3. Kowalczyk, A., Nogalski, B.: Zarządzanie wiedzą. Koncepcje i narzędzia, Difin, Warszawa (2007)
4. Przysucha, Ł.: Knowledge management in corporations – synergy between people and technology. barriers and benefits of implementation. In: Mercier-Laurent, E., Boulanger, D. (eds.) AI4KM 2017. IAICT, vol. 571, pp. 1–11. Springer, Cham (2019). https://doi.org/10.1007/978-3-030-29904-0_1
5. Domagała, P.: Internet of Things and Big Data technologies as an opportunity for organizations based on Knowledge Management. In: Proceedings of 2019 IEEE 10th International Conference on Mechanical and Intelligent Manufacturing Technologies (ICMIMT 2019), pp. 199–203. IEEE Press, Cape Town (2019)
6. Stevens, R.H.: Managing human capital: how to use knowledge management to transfer knowledge in today's multi-generational workforce. Int. Bus. Res. 3, 77 (2010)
7. https://www.shrm.org/hr-today/trends-and-forecasting/research-and-surveys/Documents/Preparing_for_an_Aging_Workforce-Oil_Gas_and_Mining_Industry_Report.pdf. Accessed 09 Aug 2021
8. Levy, M.: Knowledge retention: minimizing organizational business loss. J. Knowl. Manage. 4(15), 582–600 (2011)
9. Joe, C., Yoong, P., Patel, K.: Knowledge loss when older experts leave knowledge-intensive organisations. J. Knowl. Manage. 6(17), 913–927 (2013)
10. Alavi, M., Leidner: Review: knowledge management and knowledge management systems: conceptual foundations and research issues. MIS Q. 1(25), 107–136 (2001)
11. Román-Velázquez, J.: An empiric study of organizational culture types and their relationship with the success of a knowledge management system and the flow of knowledge in the U.S. Government and nonprofit sectors. In: Creating the Discipline of Knowledge Management the Latest in University Research, p. 77. Elsevier (2005)
12. King, W.R.: Knowledge transfer. In: Encyclopedia of Knowledge Management, p. 538. Idea Work Group Reference, United Kingdom (2006)
13. Nonaka, I.: A Dynamic theory of organizational creation. Organ. Sci. 1(5), 19 (1994)
14. Pfeiffer, J., Sutton, R.I.: The Knowing-Doing Gap: How Smart Companies Turn Knowledge into Action. Harvard Business School Press, Boston (2000)
15. Ranjbarfard, M., Aghdasi, M., López-Sáez, P., Emilio Navas López, J.: The barriers of knowledge generation, storage, distribution and application that impede learning in gas and petroleum companies. J. Knowl. Manage. 3(18), 494–522 (2014)

Characterization of Residential Electricity Customers via Deep Ensemble Learning

Weixuan Lin[ID] and Di Wu[(✉)][ID]

Department of Electrical and Computer Engineering, McGill University,
3480 Rue University, Montreal, QC, Canada
di.wu5@mail.mcgill.ca

Abstract. The household characteristics in an electric grid include the socio-economic status of households, the dwelling properties, the information on the appliance stock, and so forth. These characteristics are significantly beneficial to electric retailers, because they can be utilized to provide personalized services, improve the demand response, and make better energy efficiency programs. However, these privacy-sensitive characteristics (e.g., employment, income, age of residents) require time-consuming surveys. Also, it is difficult to gather such residential household information in a large scale. In recent years, the increasing availability of electricity consumption data makes it possible to infer household characteristics from residential electricity consumption data. A number of supervised learning methods have been proposed. Among these solutions, features are extracted from the electricity consumption patterns, and the selected features are used to train a classifier or regressor. However, the existed methods depend on a single contributing model, which can be possibly undertrained. To achieve the optimal performance of classifiers for characteristics identification, we propose an ensemble framework based on bagging algorithms. With the proposed ensemble framework, the performance of characteristic identification has been improved.

Keywords: Ensemble learning · Supervised classification

1 Introduction

The modern power grids are now facing increasing uncertainties from both the energy consumption side and the energy generation side [6,8,19]. Customer identification is of benefit for utility companies to optimize their energy programs. With detailed knowledge socio-economic characteristics of residential households, tailor saving advice can be customized for specific addressees (such as family with children or retirees). Thus, a comprehensive knowledge on customers helps electric retailers to make reliable decisions on the targeting of demand response and energy efficiency programs [17]. However, the socio-economic information of customers is privacy-sensitive and is protected by governments, especially in European [4]. Hence, in most of the cases, the knowledge of utilities about the customers does not reach beyond the address and billing information.

© IFIP International Federation for Information Processing 2022
Published by Springer Nature Switzerland AG 2022
E. Mercier-Laurent and G. Kayakutlu (Eds.): AI4KMES 2021, IFIP AICT 637, pp. 75–86, 2022.
https://doi.org/10.1007/978-3-030-96592-1_6

In the development of modern smart grids, smart meters have been becoming widely applied in residential households [7,12,13,20,22,23]. With a large amount of fine-grained data of individual costumers collected by smart meters, a deeper insight about the consumption behaviours of customers becomes available. In recent years, machine learning has been recently utilized for power grids [21]. Using data collected from smart meters, researchers have paid attention to models that identify the socio-economic characteristics of residential customers from power consumption patterns. Beckel et al. [3,4] propose a two-stage model for customer identification: first, temporal features are extracted from the original consumption patterns; second, the extracted features are used as the input for a classifier such as support-vector machines (SVM), k-nearest neighbors algorithm (KNN). With this model, various socio-economic labels have been classified by using residential consumption data. The temporal features extracted in [4] are based on priori-knowledge regarding consumption figures, ratios, temporal properties and statistical properties. Similar works have also been reported. Wang et al. decompose load profiles into partial usage patterns by using sparse coding, and the customers are classified with SVM [18]. Zhong and Tam use discrete Fourier transform and a classification-and-regression-tree (CART) to classify customers into groups based on consumption patterns [26]. Wang et al. replace the manually feature selection in [4] by a deep convolutional neural network (CNN) [17]. And the CNN in [17] is believed to be able to automatically select features from power consumption patterns. A better performance has been reported in [17] compared to [4], because CNN can extract the highly nonlinear relationships between electricity consumption in different time steps.

However, most of existing models are only dependent on a single contributing model, which might lead to an undertrained model. To achieve a better accuracy and more stable performance of a classification model, we propose a framework based on deep ensemble learning, which combines multiple classifiers to boost the accuracy and has showed its effectiveness on different types of applications [9]. This paper is organized as follow. The technical background is introduced in Sect. 2. The experiments are conducted and the results are demonstrated in Sect. 3. Last, conclusion is given in Sect. 4.

2 Technical Background

2.1 Identification of Residential Customers

To acquire socio-demographic information of residential customers, time-consuming surveys are typically required. Besides, the acquisition of some private information is strictly regulated by governments, especially in Europe. To avoid these time-consuming surveys, researchers have proposed to infer socio-demographic information based on the existing surveys and power consumption of residential households [3,4,17]. The presumption of these models is that the power consumption pattern differs between households, and is related to the socio-demographic characteristics of a household. Therefore, classifiers based on supervised learning can be built to infer socio-demographic labels from consumption patterns of a residential household.

Households with different characteristics have evident difference on their consumption patterns. Based on the Commission for Energy Regulation (CER) dataset, Fig. 1 demonstrates typical power consumption profiles of three customers. The consumption profiles of the three customers are distinct from each other. For customer #1017, most of the consumption of a day is located at the time span of $[18, 27]$ and $[31, 40]$. The consumption profile of customer #1018 has smaller peak values and less regular profiles compared to customer #1017. The consumption profile of customer #1020 demonstrates less regular pattern than customer #1017 and #1018. Table 1 shows some of the socio-demographic information of the three customers, revealing that these three customers have different living conditions as well. Hence, by comparing the daily consumption profiles in Fig. 1 and the socio-demographic information in Table 1, it is believed that the daily consumption patterns are related to the living conditions.

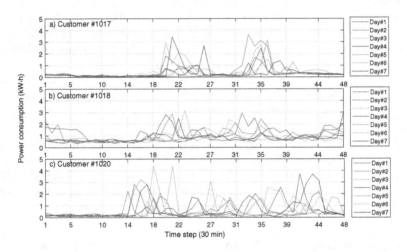

Fig. 1. Power consumption profile of a week for customers #1017, #1018, and #1020.

2.2 Problem Definition

The goal of this paper is to derive socio-demographic information from the power consumption patterns of a customer. This problem can be formulated as a supervised multi-class classification problem [17]: given the power consumption pattern $x_{i,j}$ and label $y_{i,j}$ of the i-th customer and the j-th label, a classification model $\hat{y}_{i,j} = F_j(x_{i,j}, \Theta)$ will be trained such that the loss function $L_j(\hat{y}_{i,j}, y_{i,j})$ is minimized.

Table 1. Socio-demongraphi information of customers #1017, #1018, and#1020.

Questions	# 1017	# 1018	# 1020
Age	35–65	35–65	<35
Job	Retired	Retired	Not retired
Have children?	No	No	Yes
House type	Semi-detached or terraced	Detached or bungalow	Semi-detached or terraced
Year of house	1983	1993	2000
Bedroom number	3	>5	4
Cooker type	Electric	Electric	Electric
Fllor area	<100	>100 & <200	>100 & <200

2.3 Ensemble Learning

Ensemble learning has been developed since 1990s [25]. A ensemble model is designed to combine multiple base-learners, so that the error of single base-learners can be compensated by each other. It is expected that ensemble models can outperform single base-learner [15]. Among methods in ensemble learning, Bagging and Boosting are distinct ones and commonly used.

Bagging, i.e. Bootstrap Aggregating, was first proposed in [5] and has been developed into a number of variants [2,10,11]. The essence of bagging consist of three parts: making subsets by bootstrapping on dataset, training on subsets, and aggregating single-models into the final model [15]. In this paper, we propose a ensemble framework based on bagging. Figure 2 illustrates the ensemble diagram. M subsets are bootstrapped from the training set D. For each subset D_i, a Model$_i$ is trained by a base-model. The choice of the base-models includes SVM, FNN, and LSTM. Finally, in the aggregation section, the final model is obtained by averaging the outputs from all base models.

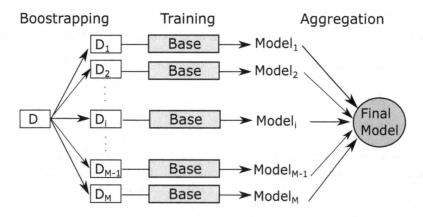

Fig. 2. Schematic of the proposed ensemble algorithm. D: training set. D_i subsets bootstrapped from the training set.

3   Experiments and Results

3.1   The CER Dataset Description

The dataset used in the experiments is provided by Commission for Energy Regulation (CER) [1], which was also used in the previous researches [4,16,17,24]. The CER dataset consists of the electricity consumption of 4232 households and a pre-trial residential survey including various questions. The power consumption spans from the 195th day to the 212th day of 2009, at a interval of 30 min. And ten of the questions as selected in [17] are the labels to be classified in the following experiments. Table 2 demonstrates the questions and the total numbers of each answer.

Table 2. Information of pre-trail socio-demographic survey in CER dataset.

Quest. no.	Socio-demographic question	Answers	Num.
300	Age of chief income earner	<65	3255
		≥65	953
310	Chief income earner is retired?	Retired	1285
		Not retired	2947
401	Social class of chief income earner	A or B or C	2482
		D or E	1593
410	Have children?	Yes	3003
		No	1129
450	House type	Detached or bungalow	1964
		semi-detached or terraced	2189
453	Age of house	>30	2161
		<30	2071
460	Number of bedrooms	≤3	2288
		≥4	1944
4704	Cooking facility type	Electrical	2960
		Not electrical	1272
4905	Energy-efficient light bulb percentage	<1/2	2041
		≥1/2	2191
6103	Floor area	<200	1438
		≥200	343

In the following experiments, the input consumption profiles have a time span of 7 days. To capture more training samples, sliding window is used as a data augmentation method. Each sampling window scans a consumption profile of 7 days (i.e. $7 \times 48 = 336$ time steps) of a customer. Adjacent windows are shifted by a time span of 7 day (i.e. 48×7 time steps), in order to pertain the weekly pattern of each customer. Each customer has ten classification labels to be classified. 4232 customers are divided into training/test set with a ratio of 0.7:0.3. It is noted that some labels are missed for certain customers. The customers with missing labels are omitted in the experiments.

3.2   Evaluation Metrics

To evaluate the performance of a classification model with M classes, the first step is to statistically obtain a $M \times M$ confusion matrix C, which is used to count the number of correct classifications and the number of misclassifications for each class [4,14,17]. Each $C_{m,n}$ denotes the number of samples of class m classified into the class n. Based on the confusion matrix, the following metrics can be derived.

Accuracy. The accuracy is defined as the rate of the samples that are correctly classified:

$$Acc = \frac{\sum_{m=1}^{M} C_{m,m}}{\sum_{m=1}^{M} \sum_{n=1}^{M} C_{m,n}} \tag{1}$$

F_1 **Score.** In a binary classification problem, the F_1 score is defined based on the number of TP (true positives, i.e. $C_{1,1}$), TN (true negatives), FP (false positives), and FN (false negatives), Pr (precision) and Re recall [14], where

$$Pr = TP/(TP + FN) \tag{2}$$
$$Re = TP/(TP + FP) \tag{3}$$
$$F_1 = 2\frac{Pr \times Re}{Pr + Re} \tag{4}$$

In the case of multi-class problem, the F_1 score is generalized to macro-F_1 score [14], which is defined as the average value of the F_1 scores of all labels, i.e.

$$F_1^{\text{macro}} = \frac{1}{M} \sum_{m=1}^{M} F_1(m) \tag{5}$$

where M is the total number of classes and $F_1(m)$ is the F_1 score of the class m.

3.3   Experimental Setups

Experiments are first conducted by various individual classifiers including biased guess (BG), Feedforward Neural Network (FNN), Long short-term memory (LSTM) based recurrent neural network, and support-vector machines (SVM). The individual model with the best performance will be chosen as the base model in the ensemble learning framework. Experiments will be conducted by the ensemble learning framework as well. To be specific, the individual models are listed as followed.

Biased Guess (BG): Knowing the proportions of different classes in the training set, the BG model assumes that all cases in the test set have the same class with the largest proportion in the training set.

FNN: A 3-layer FNN with hidden units of $(8, 8)$ are used as a classifier. The input layer has a unit number of 336, and the output layer has the unit number the same as the class number in the label.

LSTM: The LSTM baseline consists of two hidden LSTM layers with units of $(8, 8)$. The input layer has an input shape of 7×48. The unit number of the output layer is the same as the class number in the target label. And Softmax function is used for the output activation function.

SVM: The SVM baseline uses the kernel of radial basis function and the input is a time-series with 336 steps.

4   Experimental Results

4.1   Baseline Results

Table 3 demonstrates the prediction accuracy and F1 scores of the performance of baselines in test set, and Fig. 3 visualizes the accuracy and F1 results. From Table 3 and Fig. 3, it is noticed that all labels, SVM performs either better or close to the other baselines. LSTM performs close to SVM in most of the labels. And FNN is consistently the worst among the baselines. Averaging the scores in all labels, as the average accuracy shown in Table 3, SVM and LSTM have 0.669, and 0.651, respectively. While SVM and LSTM have average accuracy better than the averaged accuracy of BG (0.647), FNN has the worst average accuracy of 0.617. In terms of the average F1 socres, SVM scores 0.579, which is similar with the average F1 LSTM of 0.577. Also, FNN underperforms all baselines in average F1 scores, with 0.564.

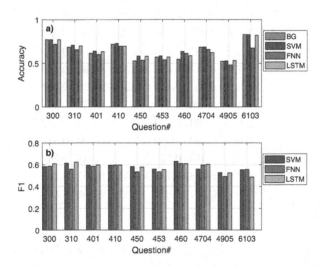

Fig. 3. Visualization of a) accuracy and b) F1 scores of baselines.

Table 3. Accuracy scores of the baseline models of SVM, FNN, and LSTM.

Questions	Accuracy				F1		
	BG	SVM	FNN	LSTM	SVM	FNN	LSTM
300	0.770	0.771	0.718	0.770	0.582	0.585	0.608
310	0.684	0.706	0.656	0.699	0.612	0.559	0.623
401	0.615	0.639	0.600	0.633	0.594	0.583	0.596
410	0.718	0.727	0.696	0.695	0.593	0.597	0.596
450	0.528	0.582	0.535	0.581	0.581	0.533	0.577
453	0.572	0.584	0.540	0.572	0.559	0.534	0.554
460	0.546	0.636	0.614	0.588	0.629	0.609	0.608
4704	0.685	0.686	0.659	0.622	0.558	0.596	0.602
4905	0.524	0.528	0.481	0.533	0.526	0.491	0.524
6103	0.829	0.827	0.673	0.820	0.553	0.556	0.487
Average.	0.647	0.669	0.617	0.651	0.579	0.564	0.577

4.2   Ensemble Model Results

Table 4. Accuracy scores of the ensemble models with base models of SVM, FNN, and LSTM.

Quest.	Accuracy			F1		
	SVM	FNN	LSTM	SVM	FNN	LSTM
300	0.775	0.766	0.694	0.593	0.618	0.589
310	0.708	0.693	0.652	0.616	0.615	0.583
401	0.643	0.625	0.576	0.595	0.590	0.555
410	0.725	0.712	0.658	0.592	0.599	0.569
450	0.574	0.562	0.571	0.569	0.559	0.566
453	0.583	0.590	0.545	0.557	0.573	0.532
460	0.635	0.624	0.554	0.630	0.618	0.562
4704	0.690	0.693	0.630	0.575	0.604	0.572
4905	0.528	0.515	0.511	0.518	0.514	0.507
6103	0.824	0.788	0.738	0.543	0.557	0.559
Avrg.	0.669	0.657	0.613	0.579	0.585	0.560

Table 4 shows the accuracy/F1 scores of the ensemble models with different base-models, and Fig. 4 visualizes the performance. From the accuracy performance shown in Fig. 4, the ensemble-SVM and ensemble-FNN model perform similarly, but ensemble-LSTM has the worst performance in most of the labels. Overall, as shown in the and average F1 in Fig. 4, ensemble-SVM and ensemble-FNN have similar average accuracy of 0.669 and 0.659, respectively. But ensemble-LSTM only has an average accuracy of 0.613, significantly smaller than the others. Regarding the average F1, ensemble-SVM, ensemble-FNN, and ensemble-LSTM have 0.579, 0.585 and 0.56, respectively. Therefore, compared to single baseline models shown in Table 3, ensemble-SVM performs similar with single-SVM, ensemble-FNN significantly outperforms single-FNN, and ensemble-LSTM underperforms single-LSTM.

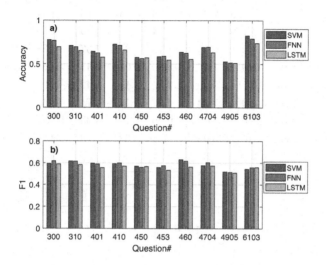

Fig. 4. Visualization of a) accuracy and b) F1 scores of ensemble models with different base-models.

Table 5 quantitatively demonstrates the improvement percentage from a single baseline model to a ensemble model, and Fig. 5 visualizes the results. In average, ensemble-SVM improves single-SVM by –0.02% in accuracy and 0.03% in F1. Ensemble-FNN improves single-FNN by 6.45% in accuracy and 3.73% in F1. And ensemble-LSTM improve single-LSTM by –5.59% in accuracy and –2.76% in F1. As shown in Table 5, we can see that with ensemble learning, the classification accuracy/F1 for FNN has been significantly improved. However, ensemble learning show little improvement on SVM, and the performance of LSTM is deteriorated after the ensemble algorithm. In the future, we will further investigate the results and try to improve the classification accuracy.

Table 5. Improvement in Accuracy/F1 scores of the ensemble models with base models of SVM, FNN, and LSTM.

Quest.	Improv. in Acc. (%)			Improv. in F1. (%)		
	SVM	FNN	LSTM	SVM	FNN	LSTM
300	0.42	6.71	−9.88	1.94	5.66	−3.13
310	0.32	5.70	−6.67	0.75	10.08	−6.52
401	0.62	4.04	−8.97	0.17	1.25	−6.82
410	−0.26	2.30	−5.33	−0.04	0.18	−4.41
450	−1.34	5.17	−1.74	−2.03	4.89	−1.89
453	−0.06	9.35	−4.78	−0.34	7.37	−3.88
460	−0.15	1.67	−5.76	0.03	1.45	−7.58
4704	0.65	5.29	1.23	3.13	1.42	−4.91
4905	0.00	7.24	−4.02	−1.53	4.71	−3.13
6103	−0.37	17.02	−9.99	−1.77	0.33	14.73
Avrg.	−0.02	6.45	−5.59	0.03	3.73	−2.76

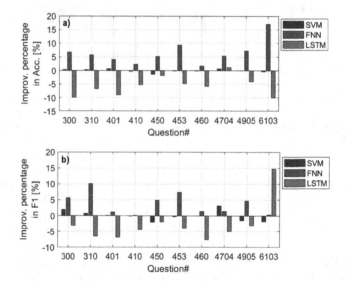

Fig. 5. Visualization of the improvement percentage in a) accuracy and b) F1 scores of ensemble models with base models of SVM, FNN, and LSTM.

5   Conclusion

Customer identification is of significant importance for utility companies to design efficient demand response program. Ensemble learning has shown to be effective for different types of applications. In this paper, we evaluated the performance of ensemble learning with different types of classification models on residential electric customer identification on a real-world dataset. Bagging method has been used to ensemble homogeneous base-models. Experiments have been conducted with base-models of SVM, FNN, and LSTM. Comparison has made between the ensemble models and single models. Results show that Bagging improves the performance of FNN the most, SVM remains the same performance after ensemble, and the performance of LSTM is deteriorated.

References

1. Commission for Energy Regulation (CER): CER smart metering project - electricity customer behaviour trial, 2009–2010 [dataset] (2012), 1st edn. Irish Social Science Data Archive. SN: 0012–00 https://www.ucd.ie/issda/data/commissionforenergyregulationcer/
2. Bauer, E., Kohavi, R.: An empirical comparison of voting classification algorithms: bagging, boosting, and variants. Mach. Learn. **36**(1), 105–139 (1999)
3. Beckel, C., Sadamori, L., Santini, S.: Automatic socio-economic classification of households using electricity consumption data. In: Proceedings of the Fourth International Conference on Future Energy Systems, pp. 75–86 (2013)
4. Beckel, C., Sadamori, L., Staake, T., Santini, S.: Revealing household characteristics from smart meter data. Energy **78**, 397–410 (2014)
5. Breiman, L.: Bagging predictors. Machine learning **24**(2), 123–140 (1996)
6. Dang, Q., Wu, D., Boulet, B.: An advanced framework for electric vehicles interaction with distribution grids based on q-learning. In: 2019 IEEE Energy Conversion Congress and Exposition (ECCE), pp. 3491–3495. IEEE (2019)
7. Dang, Q., Wu, D., Boulet, B.: A q-learning based charging scheduling scheme for electric vehicles. In: 2019 IEEE Transportation Electrification Conference and Expo (ITEC). pp. 1–5. IEEE (2019)
8. Dang, Q., Wu, D., Boulet, B.: EV charging management with ANN-based electricity price forecasting. In: 2020 IEEE Transportation Electrification Conference & Expo (ITEC), pp. 626–630. IEEE (2020)
9. Huang, X., Wu, D., Boulet, B.: Ensemble learning for charging load forecasting of electric vehicle charging stations. In: 2020 IEEE Electric Power and Energy Conference (EPEC), pp. 1–5. IEEE (2020)
10. Jiang, T., Li, J., Zheng, Y., Sun, C.: Improved bagging algorithm for pattern recognition in uhf signals of partial discharges. Energies **4**(7), 1087–1101 (2011)
11. Kuncheva, L.I.: Combining pattern classifiers: methods and algorithms. John Wiley & Sons, New York (2014)
12. Lin, W., Wu, D.: Residential electric load forecasting via attentive transfer of graph neural networks. In: IJCAI, pp. 2716–2722. ijcai.org (2021)
13. Lin, W., Wu, D., Boulet, B.: Spatial-temporal residential short-term load forecasting via graph neural networks. IEEE Trans. Smart Grid **12**(6), 5373–5384 (2021)
14. Opitz, J., Burst, S.: Macro f1 and macro f1. arXiv preprint arXiv:1911.03347 (2019)
15. Sagi, O., Rokach, L.: Ensemble learning: a survey. Wiley Interdiscip. Rev. Data Mining Knowl. Discov **8**(4), e1249 (2018)
16. Wang, Y., Bennani, I.L., Liu, X., Sun, M., Zhou, Y.: Electricity consumer characteristics identification: a federated learning approach. IEEE Trans. Smart Grid **12**, 3637–3647 (2021)
17. Wang, Y., Chen, Q., Gan, D., Yang, J., Kirschen, D.S., Kang, C.: Deep learning-based socio-demographic information identification from smart meter data. IEEE Trans. Smart Grid **10**(3), 2593–2602 (2018)

18. Wang, Y., Chen, Q., Kang, C., Xia, Q., Luo, M.: Sparse and redundant representation-based smart meter data compression and pattern extraction. IEEE Trans. Power Syst. **32**(3), 2142–2151 (2016)
19. Wu, D.: Machine Learning Algorithms and Applications for Sustainable Smart Grid. McGill University, Montreal (2018)
20. Wu, D., Wang, B., Precup, D., Boulet, B.: Boosting based multiple kernel learning and transfer regression for electricity load forecasting. In: Altun, Y. et al. (eds.) Machine Learning and Knowledge Discovery in Databases. ECML PKDD 2017. LNCS, vol. 10536, pp. 39–51. Springer, Cham (2017). https://doi.org/10.1007/978-3-319-71273-4_4
21. Wu, D., Wang, B., Precup, D., Boulet, B.: Multiple kernel learning-based transfer regression for electric load forecasting. IEEE Trans. Smart Grid **11**(2), 1183–1192 (2019)
22. Wu, D., Zeng, H., Boulet, B.: Neighborhood level network aware electric vehicle charging management with mixed control strategy. In: 2014 IEEE International Electric Vehicle Conference (IEVC), pp. 1–7. IEEE (2014)
23. Wu, D., Zeng, H., Lu, C., Boulet, B.: Two-stage energy management for office buildings with workplace EV charging and renewable energy. IEEE Trans. Transp. Electr. **3**(1), 225–237 (2017)
24. Yan, S., et al.: Time-frequency feature combination based household characteristic identification approach using smart meter data. IEEE Trans. Ind. Appl. **56**(3), 2251–2262 (2020)
25. Zhang, C., Ma, Y.: Ensemble Machine Learning: Methods and Applications. Springer, Cham (2012)
26. Zhong, S., Tam, K.S.: Hierarchical classification of load profiles based on their characteristic attributes in frequency domain. IEEE Trans. Power Syst. **30**(5), 2434–2441 (2014)

Grid Imbalance Prediction Using Particle Swarm Optimization and Neural Networks

Eren Deliaslan[1]([⊠]) [iD], Denizhan Guven[2] [iD], Mehmet Özgür Kayalica[1] [iD],
and M. Berker Yurtseven[1] [iD]

[1] Energy Institute, Istanbul Technical University, 34467 Istanbul, Turkey
deliaslan19@itu.edu.tr
[2] Eurasia Institute of Earth Sciences, Istanbul Technical University, 34467 Istanbul, Turkey

Abstract. Fluctuations in the power demand amounts, supply problems, uncertainty in weather conditions are known to cause power deviations in the real-time power market. The imbalance costs are reflected in the consumer prices in the partly liberated markets of the developing countries. Thus, the accurate short-run forecast of the electricity market trends is beneficial for both the suppliers and the utility companies to constitute a balance between the physical energy supply and commercial revenue. When both day-ahead market and intra-day market exist to respond to the power demand, forecasting the imbalances lead both the suppliers and the regulators. This study aims to optimize the grid imbalance volume prediction by integrating the Particle Swarm Optimization (PSO) and Long Short-Term Memory Recurrent Neural Networks (LSTM). The model is applied for 1 h, 4-h, 8-h, 12-h and 24-h ahead. The Mean Absolute Percentage Error (MAPE) is also calculated. As a result, The MAPE levels are found to be 27.41 for 24 h, 25.66 for 12 h, 26.77 for 8 h, 25.39 for 4 h, 9.25 for 1 h. Although improvements are foreseen both in the model and data, achievements of this study would reduce the imbalance penalties for the power generators, whereas, the regulators will organize the outages with a precise approach. Hence, the economic benefits will affect the trading prices in the long term.

Keywords: Energy market balancing · Turkish power market · Particle swarm optimization and long short-term memory

1 Introduction

After the mid-1980s, countries have engaged in the energy market reform initiatives, such as liberalization, privatization, and renewable energy technologies. In the power market, it is essential to set a balance between the energy supply and the energy consumption at any given point in time. However, electricity markets may not always ensure this balance because of the uncertainties in both supply and demand. Furthermore, due to the increasing attention on climate change and global warming, it is aimed to increase the share of Renewable Energy Sources (RES) in the energy mix. Despite the environmental

© IFIP International Federation for Information Processing 2022
Published by Springer Nature Switzerland AG 2022
E. Mercier-Laurent and G. Kayakutlu (Eds.): AI4KMES 2021, IFIP AICT 637, pp. 87–101, 2022.
https://doi.org/10.1007/978-3-030-96592-1_7

benefits of RES, there are also some concerns about their negative impacts on the electricity markets [1]. Since the electricity generation from RES is dependent on weather conditions, the amount of electricity generation becomes more unpredictable and thus, grid imbalances become much bigger. To minimize or prevent the grid unstabilities, the grid imbalances are to be compensated by the Transmission System Operator (TSO) [2]. In this context, Ancillary Services (AS), such as voltage and frequency control are critical to ensure a stable electricity network throughout the transition towards cleaner electricity production technologies [3].

Due to the transition for cleaner electricity generation, both state and private sector investments are continuously growing in Turkey. Thus, the Turkish energy market, which is one of the fastest-growing power markets all over the world, is living through a continuous change [4]. In the new market structure, the intra-day market acts as a balancing mechanism between the day-ahead market and the balancing power market. The Balancing Power Market (BPM) is conducted under the control of the Turkish Electricity Transmission Company (TETC). Although a well-balanced market is exhibited to the system operator (National Load Dispatch Centre-NLDC) within the day-ahead market, deviations arise in real-time.

The objective of this study is both to optimize the grid imbalance prediction by integrating Particle Swarm Optimization (PSO) and Recurrent Neural Networks (RNN) and to try to find an answer to "Is it possible to predict system direction?". While the PSO technique is a meta-heuristic search method whose mechanics are inspired by the swarming and collaborative behavior of biological populations, RNN is a feed-forward neural network with internal memory. RNN takes into account the current input and output that it has learned from the previous input to make a decision. The first stage is to select the best regressors among the 45 influencers of the day-ahead balancing market using the PSO technique. Following this stage, an LSTM model is utilized to predict the grid imbalance volumes for different forecast horizons.

Predicting the system direction is crucial since the system regulators reflect any imbalance costs to the utility companies which affects the consumer prices. In a way, having the information in advance is highly beneficial for the suppliers. Moreover, market participants minimize how much they lose based on the position in the market with better-predicted values. However, in the literature, there is a very limited number of studies that use the Swarm Optimization (SO) algorithm, and no other study that combines these two methods for the electricity market. This article tries to bring a new perspective for the balancing markets of the developing countries using a novel approach.

This paper is organized as follows. Grid imbalance studies have been reviewed for both modeling and forecasting approaches in Sect. 2. Section 3 presents brief information about the Turkish electricity market. In the fourth section, the methodology of the study is explained. In the fifth section, results are discussed. Finally, we conclude.

2 Literature

Power imbalance prediction is an up-to-date research topic covered in the literature for various power markets in the world. In this context, Sirin and Yilmaz [5] focused on the impact of renewable energy technologies on the Turkish balancing market. They took the generation and market outlook, wholesale electricity market structure and renewable energy policies into consideration. The results of the Quantile Regression model and Ordered Logistic Regression model showed that the system marginal price is higher in the balancing market due to the merit order effect and increase in positive balance. Using the Generalized Additive model, Soini [6] approached the subject with a similar perspective by estimating the effect of wind surplus on the price of balancing power for Denmark. The study presented that balancing power prices are consistently higher during times of lower-than-expected wind power production even after controlling for other factors. Furthermore, Gebrekiros and Doorman [7] discussed the balancing energy market clearance from the perspective of the balancing service providers and transmission system operators. In their case study, an energy balancing market design and formulation were made where reserve capacity is procured by the transmission system operators. A decrease is observed in the balancing costs and the total net imbalances are obtained for the arrangement where transmission capacity is optimally allocated.

In the literature, there are also a number of studies that model and explain the fundamentals of energy imbalance considering liberalization and renewable energy technologies. Müsgens et al. [8] analyzed the economic fundamentals that govern market design and behavior in the German Balancing power market. In a similar study, Hirth and Ziegenhagen [9] reviewed three channels (generators, policies and market to be designed, imbalance price) through which renewable energy sources and balancing systems interact. Knaut et al. [10] focused on the German balancing market by exploring the impact of the tender frequency on the market concentration. More recently, Schillinger [3] investigated the Swiss electricity balancing market and its adaptation to the energy transition.

Besides these fundamental studies, several machine learning and artificial intelligence techniques are applied in the studies that are related to the imbalance market. Kolmek and Navruz [11] forecasted the day-ahead price in the electricity balancing and settlement market for Turkey. They used ANN and ARIMA models to make a comparison according to MAPE results. In another study, Dinler [12] applied LSTM (Long Short Term Memory) method for the Turkish balancing market to test reducing the annual imbalance cost of a wind producer using the information extracted from the market data. This study revealed that the strategy performs reliably well and it may provide between a 6.3% and 11.2% decrease in the balancing cost for four tested wind power plants. In another price forecasting study, Lucas et al. [13] tried to understand the dynamics and direction of the price by using three different machine learning algorithms (Random Forest (RF), Gradient Boosting (GB), Extreme Gradient Boosting (XGBoost)) According to the results of the comparison of the model, XGBoost presented better performance and it was selected for the implementation of the real-time forecast step. The model returns a 7.89 £/MWh MAE (Mean Absolute Error), an R^2 score of 76.8% and a 124.74 MSE (Mean Squared Error).

Although there are some studies in the literature where Artificial Intelligence (AI) is used for energy imbalance, the number of these studies are quite inadequate. This study focuses on filling this gap by following the steps of Guven et al. [4]. The study aimed at predicting the sign trends in the power market by selecting the influencing factors. Genetic Algorithm (GA) with Akaike Information Criteria (AIC) was used to choose the factors with the highest impact which were then used as inputs of a Recursive Neural Network (RNN) model for forecasting the deviation model. The model was applied to perform a prediction for the day-ahead, ten hours, five hours ahead, two hours ahead and one hour ahead, and then the results were compared with other methods (Flipping coin, linear regression, GA and long term average). Although the day-ahead performance was not the best, the recommended model gave the best result among the other methods for the remaining hours.

3   The Turkish Power Market

The Turkish power market has undergone crucial changes over the last two decades. The size of the energy market has increased significantly with the changing energy resources with renewable energy technologies, increasing competition with liberalization, and the participation of numerous producers, wholesalers, retailers and regulatory entities. The value chain of the Turkish electricity market consists of [14];

- Generation
- Transmission
- Wholesale
- Distribution
- Retail
- Import Export

Private and state-owned companies are responsible for wholesale activities. Electricity is traded by the system participants in the wholesale market. The wholesale market structure is shown in Fig. 1. Electricity is traded physically and non-physically. Physical electricity trading is done through bi-literal agreements in the derivative market, as spot day-ahead and intraday electricity market and in real-time balancing and ancillary service markets. Day-ahead is an electric market where electricity is traded for the next day. In this market, the Market Clearing Price (MCP) is defined by Merit Order. In intraday, electricity transactions are organized until the market closes during the day. This market provides additional trade opportunities, liquidity, elimination imbalance etc.

In the Balancing Power Market, electricity as a requisite commodity is traded based on the market rules with spot and derivative as seen below. For the sake of the electricity grid, it is crucial to set a balance between generation and consumption.

Setting a perfect balance between generation and consumption is not easy because of the consumption forecast errors, deviations of renewable power plants and power plants failures. The balancing power market is operated by the Turkish Electricity Transmission Company (TETC, TEİAŞ). Although National Load Dispatch Centre presents balanced production and consumption with the day-ahead market, deviations may occur in real-time [4]. The results of these deviations either cause an unexpectedly insufficient supply, which makes the frequency drop below 50 Hz, or an oversupply of electricity, which makes frequency rise above 50 Hz. In the case of insufficient supply, positive balancing power is required. This can be provided by the supply side in the form of an extra amount of generated electricity or by the demand side in the form of reduced consumption. In the case of oversupply of electricity, negative balancing power has to be provided [8].

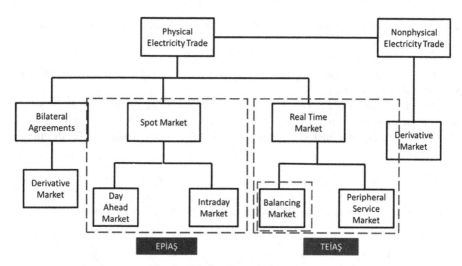

Fig. 1. Wholesale market structure

Balancing power could be provided either as a decentralized system by the responsible balancing group operators or as a centralized system service by the system operators. This is usually done for several reasons;

- In the short run, the price elasticity of both demand and supply on electricity markets is close to zero.
- The difference between the current generation and current consumption causes the frequency to deviate from the target value of 50 Hz. In the case of a decentralized balancing power system, the costs of reliability of supply would be individualized. However, the costs of an insufficient balancing power provision caused by frequency fluctuations and blackouts would be borne by all grid users.
- In the case of a centralized balancing mechanism, compensations between different balancing groups can be utilized.

4 Methodology

This study combines two methods. The first one is the Particle Swarm Optimization algorithm. It is applied to raw data to select the most effective features. The selected features are used as inputs for the deep learning algorithm.

PSO is an optimization technique inspired by the social behavior of birds or schools of fish. PSO is a population-based stochastic method and very similar to evolutionary computation techniques, such as Genetic Algorithm (GA). However, unlike GA, PSO has no evolution operator [15]. PSO has some advantages compared with other methods;

- PSO can have better results in a faster, cheaper way,
- PSO does not require the problem to be differentiable,
- PSO has very few hyperparameters,

- PSO will work on a very wide variety of tasks, which makes it a very powerful and flexible algorithm.

  In PSO algorithm there are some important features, these are; particle, velocity, swarm and optimization;

  Particle. Particles are potential solutions and they all scatter around in the search space by randomized function. In the search space, there is only one global optimum. Particles try to achieve that solution by fitness value that is evaluated by the fitness function.

$$P_i^t = \left[X_{0,i}^t, X_{1,i}^t, X_{2,i}^t, X_{3,i}^t, \ldots, X_{n,i}^t \right] \tag{1}$$

Randomly defined particles are illustrated in a basic sphere function in Fig. 2.

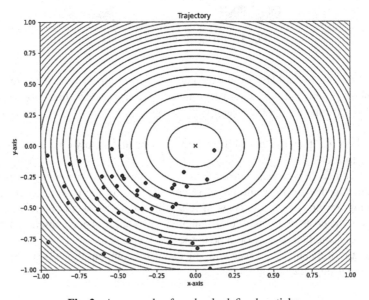

Fig. 2. An example of randomly defined particles

A fitness function is a particular type of "objective function", it specifies how close is a given design solution for achieving the set aims ($\alpha = 0.99$, $\beta = 1 - \alpha$).

$$Fitness = \alpha \times (1 - kNN_{acc}) + \beta * \left(\frac{num\ of\ feats}{max\ features} \right) \tag{2}$$

Velocity. In the search space, these particles are in the movement with a velocity allowing them to update their position over the iterations to find the global optimum. The velocity vectors are also randomized by a random function.

$$V_i^t = \left[v_{0,i}^t, v_{1,i}^t, v_{2,i}^t, v_{3,i}^t, \ldots, v_{n,i}^t \right] \tag{3}$$

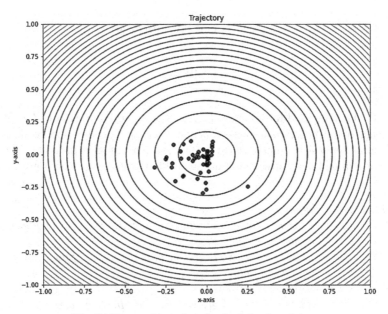

Fig. 3. New position of randomly defined particles.

New positions of randomly defined particles, which are changed by velocity vector, are illustrated in Fig. 3.

Swarm. Over the iterations in the search space, the current position is changed with the velocity. Each particle is stochastically accelerated towards its best position. This new position is called personal best and it is constantly upgraded over the iterations. These particles also accelerate towards the best solution of the swarm and are called the global best. The velocity is subject to inertia and is governed by the two best values found so far.

$$P_i^{t+1} = P_{i,}^t + V_i^{t+1} \tag{4}$$

$$V_i^{t+1} = w.v_i^t + c_1.r_1.(P_{best,personal,i}^t) + c_1.r_1.(P_{best,global,i}^t) \tag{5}$$

The first value is the best personal solution that is found by each particle. The second one is the best global solution that the swarm of particles are found so far. Thus, each particle has the best personal solution and the best global solution in their memory.

Optimization. Personal acceleration and social acceleration are stochastically adjusted by the weights r_1 and r_2. These two weights r_1 and r_2 are unique for each particle and each iteration. Hyperparameter w allows defining the ability of the swarm to change its direction. The particles have inertia proportional to this coefficient w. The inertia weight w makes a balance between the exploration and the exploitation of the best solutions found so far. Exploitation is the ability of particles to target the best solutions found. Exploration is the ability of particles to evaluate the entire research space.

$$w \in R^+; \quad \text{Inertia} \tag{6}$$

$$r_1 \in [0, 2] \quad \text{cognitive (personal)}$$
$$c_1 \in R^+ \tag{7}$$

$$r_2 \in [0, 2]$$
$$c_2 \in R^+ \tag{8}$$

The PSO can be used for feature selection. The algorithm tries to find a subset of features that optimize a certain fitness function. The feature columns in the dataset are considered as dimensions from which the PSO algorithm will select the optimal subset of dimensions through iterations. The algorithm of PSO Feature selection is given below;

Initialize Population
While (number of generation or stopping the stopping criterion is not met)
        for p =1 to number of particles
                if the fitness of Xp is greater than the fitness of pbest
                then update pbest =Xp
        for k ∈ Neighbourhood of Xp
                if the fitness of Xk greater than the fitness of gbest
                then update gbest =Xk
Next k
For each dimension d
$$V_i^{t+1} = w.v_i^t + c_1.r_1.(P_{best,personal,i}^t) + c_1.r_1.(P_{best,global,i}^t)$$
        if V_i^{t+1} is not in (Vmin, Vmax) then
        $$V_i^{t+1} = max(min(Vmax, V_i^{t+1}), Vmin)$$
        Xpd = Xpd +Vpd
        Next d
Next p
Train Neural Network with the Selected Features;

In the fitness function, firstly, we need to find the error function. For this process, k-nearest neighbors (k-NN) and multi-linear regression (MLR) methods are tested. Although both methods perform very similar, k-NN is selected to find the error value of the fitness function due to the shorter working time.

After the feature selection process is completed, the new dataset obtained with the selected feature is used as an input for a neural network. Recurrent Neural Networks is chosen because it has a memory state that processes a variable-length sequence of inputs, a quicker convergence, and more accurate mapping capability [16]. In this study, Long Short Term Memory (LSTM) as RNN variants has been chosen. LSTM has a unique feedback connection, which makes it different from the traditional RNNs. This property enables to process the entire sequence of data without treating each data point as independent. LSTM consists of cells with a gate system instead of nodes. In Fig. 4, an LSTM cell has been illustrated.

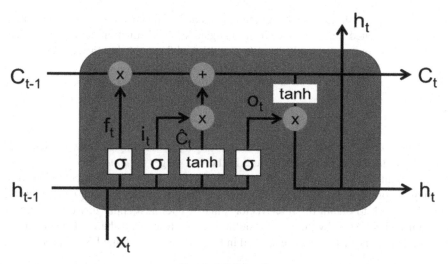

Fig. 4. LSTM cell

i_t: input gate of the cell,
f_t: forget gate of the cell,
o_t: output gate of the cell,
$\hat{C}_t$: candidate hidden state for the cell at timestamp (t)
C_t: cell state (internal memory) at timestamp (t)
h_t: predicted output from the current block.
W: the recurrent connection between the previous and current hidden layers.
U: the weight matrix that connects the inputs to the hidden layer.

The output of LSTM depends on three things;

- Cell State: The current long-term memory of the network
- Hidden State: The output of the previous point in time
- The input data at the current time step

The gate mechanism of LSTM controls the information in a sequence of the data that comes, stored and leaves the network.

Forget Gate. In this gate, previous hidden state data and the new input data are fed into the network. This network generates a vector which is an element in the interval [0, 1] by the sigmoid function. Data points that are close to 0 can be forgotten because of the less influence on the following step and the data points that are close to 1 are relevant. These outputs are then sent to pointwise multiplication.

$$f_t = \sigma(x_t U_f + h_{t-1} W_f) \tag{9}$$

Input Gate. In this gate, the goal is to determine what new information should be added to the network long term memory. The memory network has a *tanh* activation step because derivative of tanh does not reach 0 immediately; the network can learn how to combine the pre-hidden state and the new memory update vector. The input gate is a sigmoid activated network as well. In this gate sigmoid function act as a filter. The output of the input gate is again pointwise multiplied then added to the cell state.

$$i_t = \sigma(x_t U_i + h_{t-1} W_i) \tag{10}$$

$$\hat{C}_t = \tanh(x_t U_g + h_{t-1} W_g) \tag{11}$$

Output Gate. The main objective of output gate is deciding the new hidden state. The output gate works like forget gate. In the first stage, the *tanh* function is applied to the current cell state pointwise to obtain the squished cell state. This is followed by passing the previous hidden state and current input data through the sigmoid activated neural network to obtain the filter vector. Finally, one must apply this filter vector to the squished cell state by point wise multiplication. To convert the hidden state to the output, a linear layer needs to be applied in the very last step of the LSTM process.

$$o_t = \sigma(x_t U_o + h_{t-1} W_o) \tag{12}$$

$$C_t = \sigma\left(ft * C_{t-1} + i_t * \hat{C}_t\right) \tag{13}$$

$$h_t = \tanh(C_t) * o_t \tag{14}$$

The methodology of this study is illustrated in Fig. 5.

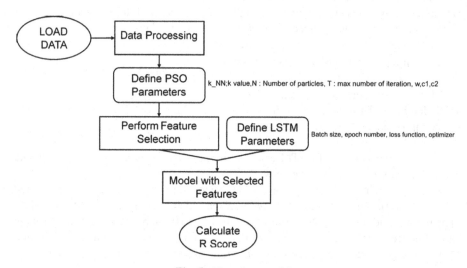

Fig. 5. Flowchart model.

Data. Market energy data from the Transparency Platform and TETC, historic data and the system direction as a target are chosen for the input for the PSO algorithm to find the best features. For market data, system marginal price (SMP) and market clearing price (MCP) for both lagged day and latest hour, frequency capacity price, petroleum pipeline company (BOTAŞ) price and net loading output data are considered. In addition, CCGT coal, renewable energy, demand forecast, and demand forecast error are used as inputs. Year, day of the year, month, day of the month, weekday and hour are the historic features of the raw data input. The hourly values of the last two years are selected for the data. The correlation matrix of the data is illustrated in Fig. 6.

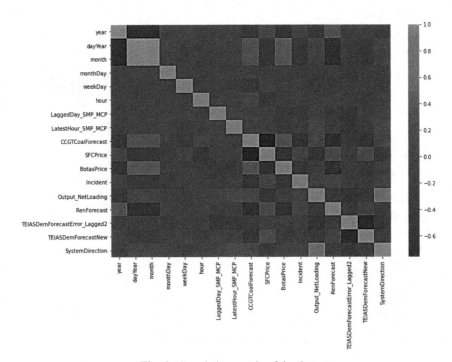

Fig. 6. Correlation matrix of the data

A total of 17 features are used as inputs for the PSO model. Best features are obtained from the model.These features are used as new parameters for the imbalance volume prediction data that will be used to find the results with the LSTM model.

5 Results and Discussion

In the first step, for the fitness function of the PSO algorithm, we need to find the error function. For this process, k-NN and the MLR are tested. k-NN provided the feature selection results quicker than the MLR (12 min vs. 17 min). k-NN is chosen for the error function (1-accuracy). For the k value, asset of 3 to 9 are tested and the best value

5 is taken for consideration. The hyperparameters are chosen from the literature; the inertia (w) is 0.9, the acceleration coefficient for personal best is 2, and the acceleration coefficient for global best is 2. System upper and lower bounds are defined as 1 and 0. Also, a total of 3 features are selected out of 17. These are latest hour SMP and MCP, BOTAŞ price and net loading output. The feature selection process is illustrated in Fig. 7.

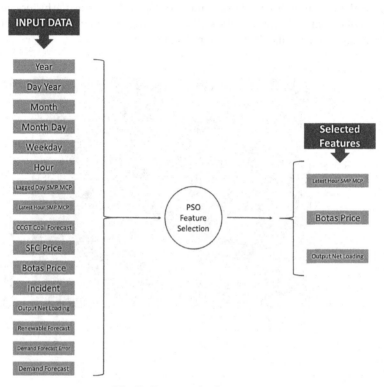

Fig. 7. Feature selection process

The change in fitness function error over iterations is given in Fig. 8. Before starting the PSO modeling, the data were normalized. While the error values were 0.094 in the first iteration, they decreased to 0.04 levels as the particles approached the global optimum at the end of each iteration and continued at a constant value after a certain point.

Selected features created a new data set with a new target which is imbalancing volume. Since we used classification methods to find fitness function error, we cannot use the imbalance volume at the feature selection process. This data set is used as input for LSTM. In the LSTM, 50–150 and 250 units are tested, but there is no significant change in the results. Eventually, 50 units are selected because of the faster performance. After a certain point, the error of the validation and training data remained stable. Hence, for the fastest performance, data are trained for 32 times. Optimizer is selected as *adam*. 32

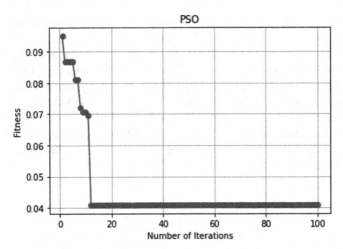

Fig. 8. Feature selection process of PSO

training examples are utilized in one iteration. Error is calculated using Mean Absolute Error and a decline is observed in each train as illustrated for 1 h ahead prediction in Fig. 9.

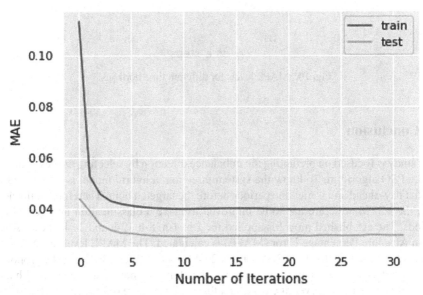

Fig. 9. Mean Absolute Error for 1 h ahead model performance

This model is applied for 1 h, 4-h, 8-h, 12-h and 24-h ahead and Mean Absolute Percentage Error (MAPE) is calculated. While the proposed model produces much better result for 1 h ahead, it may not be sufficient for 4 h and earlier predictions. However, this model is likely to give better results than many models in the literature when the 1-h-ahead predictions are compared (data usage can make differences). In Fig. 10, the change of MAPE values according to time is shown. Although the MAPE values are higher in the previous estimates, it can be said that the system user will reduce the risks in taking an early position according to the model results. The MAPE levels are found to be 27.41 for 24 h, 25.66 for 12 h, 26.77 for 8 h, 25.39 for 4 h, 9.25 for 1 h. Despite the low error rate of this model, the dataset should be reviewed carefully. The biggest problem of the data set is detected as the lack of hourly weather data which may affect the model significantly.

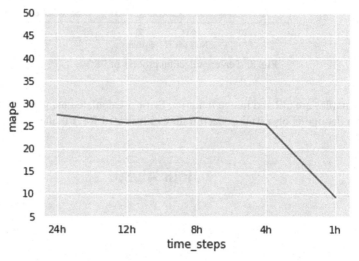

Fig. 10. MAPE levels for different time horizons

6 Conclusion

This study is focused on predicting the imbalance direction by selecting the best features by the PSO algorithm. To know the system direction, a hybrid model is constituted to predict day-ahead and some time periods before the targeted hour. Therefore, in the next step, the LSTM structure has to be improved to make a classification in several time periods. The established model is applied for 1 h, 4-h, 8-h 12-h and 24-h ahead. Also, Mean Absolute Percentage Error (MAPE) is calculated. The MAPE levels are found to be 27.4 for 24 h, 25.7 for 12 h, 26.8 for 8 h, 25.4 for 4 h, 9.3 for 1 h. While the proposed model produces a much better result for 1 h ahead, it may not be sufficient for 4 h and earlier predictions. However, it can be said that with the increasing number of artificial intelligence applications in the energy field, the imbalance prediction will get closer to real-time and the benefits of system participants will increase. Moreover, acquisitions of the established model will assist all stakeholders and key market players in countries which has partly liberated power markets.

References

1. Hache, E., Palle, A.: Renewable energy source integration into power networks, research trends and policy implications: a bibliometric and research actors survey analysis. Energy Policy **124**, 23–35 (2019)
2. Lago, J., Poplavskaya, K., Suryanarayana, G., De Schutter, B.: A market framework for grid balancing support through imbalances trading. Renew. Sustain. Energy Rev. **137**, 110467 (2021)
3. Schillinger, M.: Balancing-market design and opportunity cost: the Swiss case. Utilities Policy **64**, 101045 (2020)
4. Guven, D., Ozozen, A., Gülgün Kayakutlu, M., Kayalica, O.: Predicting power deviation in the Turkish power market based on adaptive factor impacts. In: Eunika Mercier-Laurent, M., Kayalica, Ö., Owoc, M.L. (eds.) AI4KM 2021. IAICT, vol. 614, pp. 213–234. Springer, Cham (2021). https://doi.org/10.1007/978-3-030-80847-1_14
5. Sirin, S.M., Yilmaz, B.N.: The impact of variable renewable energy technologies on electricity markets: an analysis of the Turkish balancing market. Energy Policy **151**, 112093 (2021)
6. Soini, V.: Wind power intermittency and the balancing power market: evidence from Denmark. Energy Econ. **100**, 105381 (2021)
7. Gebrekiros, Y., Doorman, G.: Balancing energy market integration in Northern Europe-modeling and case study. In: 2014 IEEE PES General Meeting Conference & Exposition, pp. 1–5. IEEE (2014)
8. Müsgens, F., Ockenfels, A., Peek, M.: Economics and design of balancing power markets in Germany. Int. J. Electr. Power Energy Syst. **55**, 392–401 (2014)
9. Hirth, L., Ziegenhagen, I.: Balancing power and variable renewables: three links. Renew. Sustain. Energy Rev. **50**, 1035–1051 (2015)
10. Knaut, A., Obermüller, F., Weiser, F.: Tender frequency and market concentration in balancing power markets (2017). Unpublished
11. Kölmek, M.A., Navruz, I.: Forecasting the day-ahead price in electricity balancing and settlement market of Turkey by using artificial neural networks. Turk. J. Electr. Eng. Comput. Sci. **23**, 841–852 (2015)
12. Dinler, A.: Reducing balancing cost of a wind power plant by deep learning in market data: a case study for Turkey. Appl. Energy **289**, 116728 (2021)
13. Lucas, A., Pegios, K., Kotsakis, E., Clarke, D.: Price forecasting for the balancing energy market using machine-learning regression. Energies **13**, 5420 (2020)
14. Overview of the Turkish Electricity Market. pwc.com.tr & Aplus Enerji, August 2020
15. Eberhart, R.C., Shi, Y.: Comparison between genetic algorithms and particle swarm optimization. In: Porto, V.W., Saravanan, N., Waagen, D., Eiben, A.E. (eds.) EP 1998. LNCS, vol. 1447, pp. 611–616. Springer, Heidelberg (1998). https://doi.org/10.1007/BFb0040812
16. Brito, R., Fong, S., Zhuang, Y., Wu, Y.: Generating neural networks with optimal features through particle swarm optimization. In: Proceedings of the International Conference on Big Data and Internet of Thing – BDIOT 2017, pp. 96–101. ACM Press (2017)

Collective Intelligence of Honey Bees for Energy and Sustainability

Mieczysław L. Owoc$^{(\boxtimes)}$ (iD)

Wrocław University of Economics and Business, 118/120 Komandorska Street, 53-345 Wrocław, Poland
mieczyslaw.owoc@ue.wroc.pl
http://www.ue.wroc.pl/en/

Abstract. From the very beginning the most promising AI methods are inspired by human environment and nature. Especially, collective intelligence of non-human societies can surprised researchers and developers of new solutions. It is matter of specific abilities of particular species oriented on cooperation but moreover awareness of precisely defined goals and resources used in very optimal ways. For example we may admire methods of building cells by honey bee society and organizing of their works; therefore for example problems of tasks planning or optimization of collection of pollen by bees are being solved through certain sort of common intuition and collective intelligence. Proposed Bees algorithm (as an example of swarm algorithms) has been applied in continuous domains (optimization of neural networks) or combinatorial ones (scheduling jobs for a machine). The goal of this paper is presentation of collective intelligence useful in relatively new directions: energy acquisition and selected processes assuring sustainable development. Both directions seem to be very innovative and promising–especially in the ecosystems context.

Keyword: Collective intelligence · Honey bees society · Sustainability · Bee algorithms

1 Introduction

For ordinary people honey bee colonies are useful insects producing honey and wax. In reality, honey bees play very important role in the nature delivering services in broadly understood ecosystem. Their activeness and products cover three main areas: *regulating, provisioning and cultural* – it is presented in Fig. 1. The provisioning services are essential for short- and long-term be society surviving while two other (regulating and culture) are closely related to the surroundings of apiary.

Regulating functions connected with pollination and biodiversity are strictly connected with interactions typical for complex ecological systems. Pollination is a process of transporting pollens among plants and therefore is inevitable for its reproduction in the vegetable world. This process depends on the symbiotic relationship between pollinator and pollinated plants and honey bees can be recognised as the major pollinator workhorse and the same are essential components of terrestrial ecosystem.

E. Mercier-Laurent and G. Kayakutlu (Eds.): AI4KMES 2021, IFIP AICT 637, pp. 102–116, 2022.
https://doi.org/10.1007/978-3-030-96592-1_8

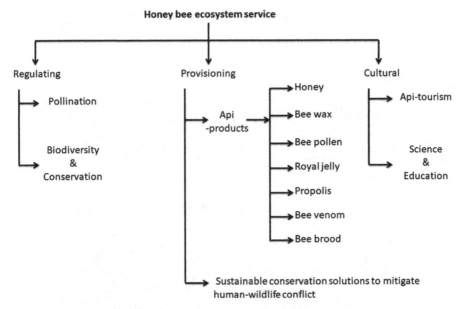

Fig. 1. Ecosystem services of honey bees. Source: [2]

Bees are the most famous for their **provisioning** function. The product of beehive which is identified with their activities is honey. This product is a result of collection nectar by honey bees and further transformation involving their digestion system using enzymes. The total production of honey is systematically growing despite of real obstacles in bee-keeping mostly coming from human side: pollution, use pesticides and the like. The same tendency can be observed in wax production (see Fig. 2).

The third significant for people product is *propolis*. This extraordinary substance consisting of hundreds ingredients honey bees colonies use to seal the hive; very important property of propolis is antibacterial feature useful in medicine. The other api products are essential for bee colonies–people very rarely can utilize them.

The last itemized function defined as **cultural** is connected with impact and interaction apiary on environment; tourism managed around honey bees societies and generally speaking gathering knowledge about live and behaviour of the presented colonies create conditions for development of apiary sector.

As we discussed earlier, all the mentioned functions constitute the services in this specific ecosystem also in terms of sustainability and conservation determining progress of our civilization. Bees are symbols of diligence and sacrifice for common colony and their care for the welfare of the hive are appreciated in the proverbs of various nations for example: as busy as a bee.

The mystery and the willing of applying some natural habits of honey bee colonies prompted many researchers to investigation of bee societies live. All these research directions can be diversified in the following way.

Firstly, an essence and different aspects of their social live were defined as an important topic of investigation; such problems like communication among bees, cooperation

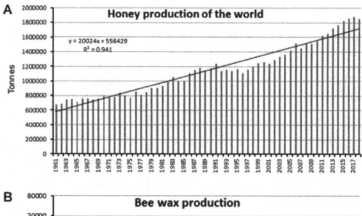

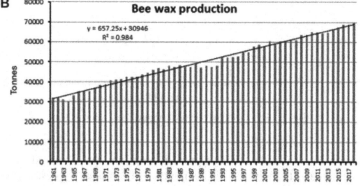

Fig. 2. World honey and wax production. Source: [24]

typical for all phases in their life and common decision taking gather scientists from many areas. This kind of research creates opportunity for considering interdisciplinary aspects of honey bees society functioning. Therefore *collective intelligence* of honey bees is critical in these research.

Secondly, big activities and diligence of honey bees can be investigated in the *energy context*. There are at least two orientations in this area. During performing many tasks inside and out of the hive honey bees must rational use their energetic resources. It can be also (at least potentially) possible to use their mobility to transform it to useful for people energy.

Thirdly, honey bee colonies can be analysed from the point of their abilities to survive. The history of honey bees societies reaches over hundreds million years so these colonies elaborated special mechanisms assuring *sustainable development* of their civilization.

All these directions of research will be discussed in the next sections.

2   Honey Bee Intelligence

One of most fascinating phenomena in our world is behavior of animal societies especially working together for their own welfare. It is possible as a result of their developed brain and abilities to cooperate in colonies. Honey bees' brains is relatively small (comparable to sesame seed) but with more neurons than insects of similar size with greater brain density (ten times denser than the brains of mammals). Their sophisticated sensory system gives them excellent sight including keen smell, taste, and touch. In addition, their brains appear to have neuroplasticity with capabilities of learning, adapting and performing the same functions.

Honey bees are *smart*–there are many observations and experiments confirming their learning and memory capabilities–see: [25]; for example they can:

- identify and remember colors and landmarks,
- distinguish among different landscapes, types of flowers, shapes, and patterns,
- remember route details up to six miles over several days,
- conceptualize a map, determine the shortest distance between two points, and take a different route for their outbound and inbound journeys,
- navigate even in the dark.

Members of honey bee society are able to take on many *different roles in their lifetime*, each requiring different skills. The youngest bees are nurse bees, tending to the brood (the pupa or larva stage). Later they make the honey combs, forming perfect hexagonal shapes. Lastly, they become foragers, finding honey and bringing it back to the hive. It is worth to stress their abilities to distinguish harmful fungi from harmless ones. They can prepare medication (mentioned earlier propolis) after discovering harmful fungi.

The most crucial ability of intelligent objects is *ability to think* – so how it looks in case of bee colonies? Experiments and observations show that honey bees do indeed have the ability to think. As it was announced - they have an ability to learn many new things very quickly (compared to other insects, they have a much greater ability to learn and remember). They are capable of abstract thought, decision-making, and planning. They also show an ability to count and an understanding of time. The foragers have to perform many tasks that require intelligence - they must find flowers, figure out if they are a good source of nectar, find their way back to the hive, and then share this information with the other foragers. They do not use tools in nature, but they can be taught to use tools. They may be capable of cultural transmission. Experiments have shown that when one bee was taught to pull a string to get a sugary reward, another bee learned the trick just by watching the first bee. Even more surprising, they could teach this trick to other bees-see: [25].

Very significant ability which assure cooperation among honey bee society members is *communication*. Research on bee language was conducted by the Nobel Prize-winning biologist Karl von Frisch (see [8]). They have a sophisticated communication system using a symbolic language. This system is not found in any other animal. They communicate using "body language" and "dance." When a honey bee finds a good source of nectar, she will return to the hive and tell the other bees where to find it by doing a dance to communicate both the location and quality of the source. Bees apply two sorts of dance in order to communicate food source and direction of where it can be found.

The simplest dance is the "round dance." The bee dances around in a circle. The same the bee tells the others that the food source is very close. Other workers can go out immediately and just sniff around to find it; they know what to sniff for because they can smell the odor on the forager's body or because she has given them a little taste. Details of this dance are presented in Fig. 3.

Fig. 3. The round dance of honey bees. **Source:** [21]

If the nectar source is further away, the forager does a "waggle dance" – presented in Fig. 4. The bee begins by walking in a straight line to indicate the direction of the food source relative to the sun. She then dances to create an angle which indicates the precise location of the food source relative to the sun. She waggles her body as she does this; the number of waggles indicate the distance to food.

Imagine a line connecting the hive and the food source, and another line connecting the hive to the spot on the horizon just beneath the sun. The angle formed by those two lines indicates the location of the food source relative to the sun.

The bee adjusts the angle of her dance to account for the fact that the sun is always moving across the sky. The forager takes this into account and changes the angle of her dance every four minutes by one degree to the west.

Finally, the time she takes doing the dance indicates the strength of the headwind. This tells the other bees how much honey they need to eat to have enough energy for the trip. There are some differences in defining distance to the food source by particular species of bees; for example changing to the tail at about 40 m for Italian and Caucasian bees. It also interesting that this changes at about 90 m for Corniolan bees. Therefore we may say about dialects in the bee language.

It is curious additional feature of bees confirming their collective intelligence, namely *decision making process* – compare [28]. Foraging requires no central decision making. Each bee knows only about her own nectar source. If the nectar source is poor, she will quickly abandon it and will not do a waggle dance to direct her hive mates to it. Instead, she will watch the waggle dance of another bee who has found a good source of nectar

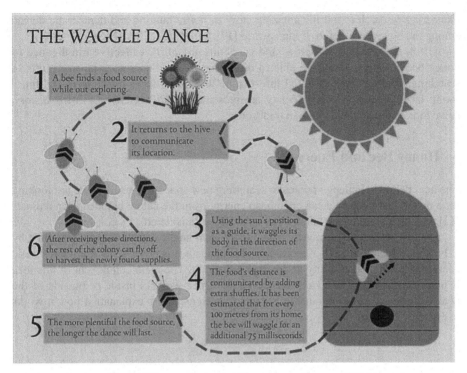

Fig. 4. The waggle dance of honey bees. Source: [22]

and turn her foraging efforts to that source. On the other hand, if her source is good, she will work that source consistently and recruit other bees to join her.

Each individual bee makes her own individual cost-benefit analysis, which affects whether or not she will abandon a source or recruit other bees to it. If she abandons her source, she will follow another bee at random; she does not compare sources. Since only bees with good sources are recruiting, the bee who has abandoned her source will almost certainly move to a better source. The result is that the entire hive has optimized foraging – the decisions of thousands of individual bees lead to optimal collection of food resources.

The value of a nectar source is not based only on the sugar content of the source—it appears that distance from the hive, weather conditions, the hive's nutritional status, the depletion of the source, and other pertinent factors are all involved.

Honeybees make decisions collectively and democratically. Every year, faced with the life-or-death problem of choosing and traveling to a new home, honeybees stake everything on a process that includes collective fact-finding, vigorous debate, and consensus building. In fact, as world-renowned animal behaviorist Thomas Seeley reveals, these incredible insects have much to teach us when it comes to collective wisdom and effective decision making. A remarkable and richly illustrated account of scientific discovery, *Honeybee Democracy* brings together, for the first time, decades of Seeley's pioneering research to tell the amazing story of house hunting and democratic debate among the honeybees before flying swarm [17].

All these presented properties and capabilities confirm **collective intelligence** of honey bee colonies. There are able to discover new food source, compare results of searching, communicate essential information and make decisions collectively. Interesting facts about honey bee society are presented in [29] referring to their history, behaviour, and impacts on human traditions.

3 Honey Bee and Energy

The next defined challenge relates to searching new sources of energy. We are looking for a method to produce clean and cheap energy from the nature but also farm animals [12] and other form of human activities should be considered.

The original proposal of acquiring energy from honey bees hives was presented by H.F. Abou-Shaara see: [1]. According to his concept it is possible to apply the kinetic energy of honey bee mobility to produce electricity. The main component of this idea is placement of special devices on the paths of bee workers inside or outside of the hive. This device working outside of hives is presented with explanation how it works in Fig. 5.

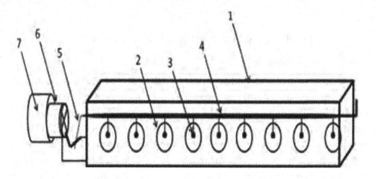

Parts of the external device working outside the hives. This device is a box (1) with dimensions (5x5x40 cm, HxLxW), this box is placed in front of the hive and contains holes (2) with a diameter of 7 mm per each, the bees can pass from these holes to push a small transparent ball (3) moving the tube (4), and then the arm (5) push the fan (6) to generate electricity from the dynamo (7). This dynamo can be used to charge batteries, and these batteries can be further used to operate various devices.

Fig. 5. External device of the hive. Source [1]

The next possible place of honey bees movement is the interior of the hive (especially workers in early stages of their live). They are employed to perform different actions connected with feeding larvas, producing wax, cleaning the hive etc. Similarly to the previous content of the device through pushing a small transparent ball is possible to generate electricity from the installed dynamo. Details of this solution is demonstrated in Fig. 6.

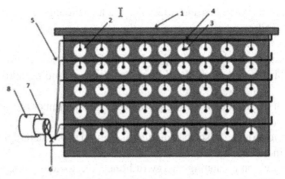

Parts of the internal device working inside the hives. This device contains the carrier frame (1) (with dimensions of 25x 42 cm L x W) which contains holes (2) with diameter of 7 mm, and the bees can pass through these holes to push small balls (3), hence moving the horizontal tube (4), and moving part (5) which connects to an arm (6) to move the fan (7) to generate electricity by the dynamo (8). This dynamo can be used to charge batteries, and these batteries can be further used to operate various devices.

Fig. 6. Internal device of the hive. Source [1]

It is also possible to place separate improved devices (as a board of many tubes with balls) in different places inside and outside of the hive and connect partial outputs. This concept is showed in Fig. 7 – such combined solution should generate more power.

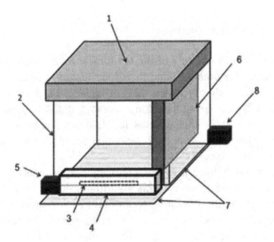

The positions of the devices in relation to the hive. The hive consists of upper cover (1), a box (2) and hive entrance (3). The external device (4) is placed in front of the entrance (3) to utilize the movements of the forager bees when entering or leaving the hive, while the fan and the dynamo (5) are placed beside the device. The internal device (6) is placed inside the hive beside the hive frames to utilize the movements of the bees inside the hive to generate electricity, while the fan and the dynamo (8) is placed outside the hive. The third device (7) is placed under the hive box while the fan and the dynamo (8) are placed outside the hive. The third device utilizes the movements of the bees inside and outside the hives because this device covers the hive bottom and the front part of the hive.

Fig. 7. Potential placement devices in the hive. Source [1]

The presented concept can be used close to existing apiaries. These devices are not expected to cause harmful impacts on the colonies or disturb daily (outside of the hive) or 24 h (inside of the hive) of bee activities. The main challenges of the proposed concept can be defined as follows:

- the solution is addressed to beekeepers to help in their work; any transmission of electricity to far located users should be considered including additional costs, see [9],
- implementation of this idea is strongly connected with number of hives in the apiary; it is rather recommended for bigger farms,
- the area of application the described devices is limited to isolated regions as additional option to obtain electricity.
- additional research on balancing energy of honey bee society in the hive is welcome.

Anyway this investigation create opportunity for a discussion on development devices or similar mechanisms based on kinetic energy. It still open consideration about other forms of obtaining energy from insects. Unfortunately, in this case collective intelligence of honey bees has no special meaning but their ability to tolerate something new and tendency to survive in the new conditions confirm adaptation features typical for smart societies. Considering bee colony activities (especially flying swarm) and selection of routes some – they are evidences confirming their orientation on energy lines (see [26] and [27]). One yet another connection between honey bees and flowers: the "electric" signals from flowers can attract foragers – see: [30].

4 Honey Bee and Sustainability

Keens of honey bees admire the role of their colonies in the nature (mentioned earlier importance of vegetable pollination) and their long history – there are some evidence their presence over 130 million year ago [23]. Thanks to their sense of responsibility for bee existence confirmed during their long history and ability to survive this society cane useful in modern time; brother Adam said: "Listen to the bees and let them guide you". Accordingly to his saying scientists elaborated bee algorithms based on behaving of foragers during vegetable pollination.

Honey bee algorithms belong to swarm intelligence algorithms in the group of insect based. The idea of the basic version of this algorithm mimics the foraging strategy of honey bees to look for the best solution to an optimisation problem. Each candidate solution is thought of as a food source (flower), and a population (colony) of n agents (bees) is used to search the solution space. Each time an artificial bee visits a flower (lands on a solution), it evaluates its profitability (fitness). Discussion on theoretical aspects, improvements and application of the bees algorithm is presented in [10].

The bees algorithm consists of an initialisation procedure and a main search cycle which is iterated for a given number T of times, or until a solution of acceptable fitness is found. Each search cycle is composed of the separate procedures. The list of steps typical for a simple bee algorithm is presented in Fig. 8.

Initialisation of the algorithm relies on defining a set of parameters like: number of scout bees (playing this following more experienced workers) and number of elite bees (discovering new sources of nectar, water etc.), number of selected regions, number of recruited bees around the regions and stopping criteria.

1. Initialise population with random solutions.

2. Evaluate fitness of the population.

3. While (stopping criteria not met) //Forming new population.

4. Select elite bees.

5. Select sites for neighbourhood search.

6. Recruit bees around selected sites and evaluate fitness.

7. Select the fittest bee from each site.

8. Assign remaining bees to search randomly and evaluate their fitness.

9. End While

Fig. 8. Pseudo code of the bee algorithm. Source: [16]

General idea of this algorithm looks similarly to evolution algorithms (a cycle and certain actions changing population with a defined criteria of fitness evaluation). The main difference refers to inclusion of discovered sources and communication about it to other workers (waggle dance) and applying two kind of searching: local and global. An example of such algorithm used for automatic design of control system for robot manipulators is presented in Fig. 9.

There are several areas of successful implementation of bee algorithms – power systems seems to be very ambitious challenge for different phases o supporting this industry. Very often the discussed algorithm can be connected with additional methods. Such approach is presented in [3] where Authors decided to include machine learning and neural networks to reactive power optimization. Details are demonstrated in Fig. 10. As previously two sorts of "objects" are employed: workers and scouts both with the defined goal as searching nectar source. According to earlier assumptions workers are oriented on local search while scouts following earlier defined strategies are allocated to global search. The algorithm is based also on knowledge learning through defined Q-value matrix where particular actions are evaluated (reward function) finally generating optimised model useful in large-scale Power Systems.

Additional examples of utilising bee algorithms in the power industry can be found in: [11, 16] and [6]. Different aspects in research of the discovered problem confirm flexibility of these algorithms.

The presented instances of bee algorithms implementation are focused on power industry but there are many others sectors where different variants of bee algorithms have been successfully applied. More synthetic view on the discussed algorithms is showed in Fig. 11.

First, *typical areas* where ABC algorithm can be useful; the range of potential application is very wide: starting from supporting computer graphics (image segmentation)

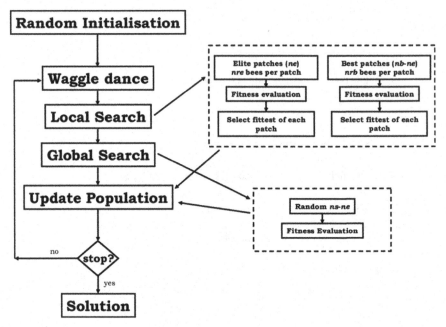

Fig. 9. Flowchart of the bee algorithm. Source [7]

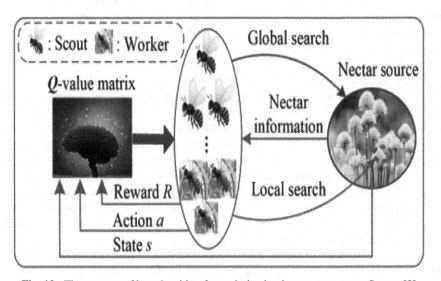

Fig. 10. The concept of bee algorithm for optimization in power systems. Source [3]

through learning tasks (training neural networks) and allocation problems (scheduling and redundancy) up to classification (patterns).

Second, basic *control parameters* essential in experiments are stressed. Apart of number of onlooker bees, food sources and limits in cycles can be determined.

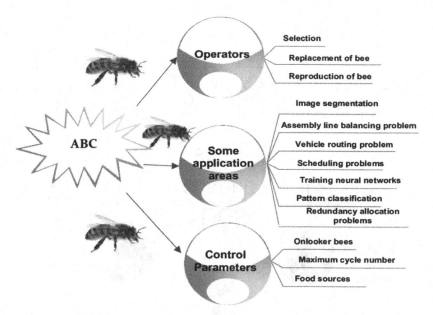

Fig. 11. Artificial Bee Colony (ABC) algorithm. Source [4]

Third, standard *operators* are stressed. The most adequate in ABC algorithm are: selection of bees, replacement some of them and reproduction.

Summing up, applications and variants of bee algorithms (additionally: BeeHive, Bee Colony Optimization, Bee Swarm Optimization, Virtual Bee Algorithm, Marriage Bee Optimization) create opportunities for applying them in different sectors, not only for optimization purposes. This means real impact on sustainable development and stability of systems (compare to [16]).

The problem of meaning bees for achieving sustainable development is discussed in [13]. Main dimensions of supporting SDGs (Sustainable Development Goals) are depicted in Fig. 12.

Findings of the Authors in [13] can be formulated as follows:

" A holistic view of ecosystems including wild and managed bees and humans is necessary to address sustainability challenges. By employing a system approach, we can better understand the interconnections between elements within coupled human–environment systems. We strongly advocate the need for appropriate natural resource management approaches for maintaining sustainable systems as vital for allowing the continued success of bees in their natural role.

We summarise our findings by suggesting eight key thematic priority areas whereby bees can play a crucial role in meeting the SDGs."

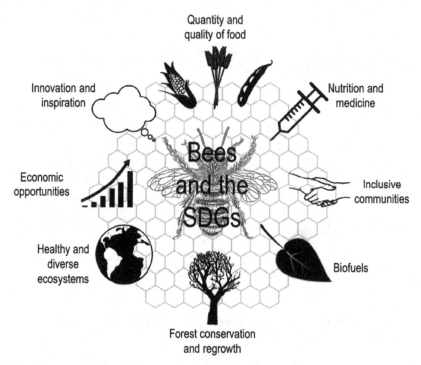

Fig. 12. Bee colonies and SDGs. Source: [13]

5 Conclusions

Nowadays, people more or less are engaged in saving our world. One of the most important problems is ensuring a balance in nature. The role of honey bees in this process is very significant. Final conclusions are:

1. Honey bee intelligence considered as collective intelligence is crucial in adaptation for changing conditions of colonies life.
2. Ecological aspects of honey bee life are extremely important for their survive but also for keeping a balance in nature.
3. Original approaches referring to looking for alternative sources of energy embrace honey bee society; the presented devices can be useful in supporting in very limited range bee-keepers.
4. Honey bee intelligence was inspiration for the new swarm algorithm. Implementation of different bee algorithms in different areas confirms their universal and flexible character.

Further research can be devoted for presentation experiments and making different types of decisions based on collective intelligence and bees democracy following G.B. Shaw sentence: "Go to the bee, thou poet: consider her ways and be wise" quoting from [17].

References

1. Abou-Shara, H.F.: Devices to generate clean and renewable energy from honey bee hives. Arthropods **8**(3), 97–101 (2019). www.iaees.org
2. Aryal, S., Ghosh, S., Jung, C.: Ecosystem services of honey bees; regulating, provisioning and cultural functions. J. Apiculture **35**(2), 119–128 (2020)
3. Cao, H., Yu, T., Zhang, X, Yang, B., Wu, Y.: Reactive power optimization of large scale power systems: a transfer bee optimizer application. Processes **7**, 321 (2019). https://doi.org/10.3390/pr7060321, www.mdpi.com/journal/processes
4. Chinnasamy, R., Kandasamy, G.: Investigation on bio-inspired population based metaheuristic algorithms for optimization problems in ad hoc networks. Int. J. Math. Comput. Nat. Phys. Eng. **9**(3), 163–170 (2015)
5. Crawford, B., Soto, R., Cuesta, R., Paredes, F.: Application of the artificial bee colony algorithm for solving the set covering problem. Sci. World J. **2014**, 1–8. Article ID 189164 (2014)
6. Dinh, L.L., Ngoc, D.V., Vasant, P.: Artificial bee colony algorithm for solving optimal power flow problem. Sci. World J. **2013**. Article ID 159040 (2015)
7. Fahmy, A.A., Kalyoncu, M., Caselani, M.: Automatic design of control systems for robot manipulators using the bees algorithm. J. Theoret. Appl. Inf. Technol. **49**(1) (2013)
8. von Frisch, K.: Bees: Their Vision. Comstock Publishing Associates, New York (2014)
9. Hadjur, H., Ammar, D., Lafevre, L.: Analysis of Energy Consumptions in a Precision Beekeeping System. In: IoT 2020: Proceedings of the 10th International Conference on the Internet of Things, October 2020. Article No. 20 (2020). https://doi.org/10.1145/3410992.3411010
10. Koc, E.: The Bees Algorithm Theory, Improvements and Applications. Ph.D. thesis at University of Wales, Cardiff (2010)
11. Kokin, S., Manusov, V., Matrenin, P.: Optimal placement of reactive power sources in power supply systems using particle swarm optimization and artificial bees colony optimization. In: IEEE Conference Paper (2018). https://www.researchgate.net/publication/318326509
12. Paras, S.V.K., Chaudhary, A.: Generation of electricity by utilization of power of draught animal. Indian Res. J. Extension Educ. **12**(2), 150–153 (2016)
13. Patel, V., Pauli, N., Biggs, E., Barbour, L., Boruff, B.: Why bees are critical for achieving sustainable development. Ambio **50**(1), 49–59 (2020). https://doi.org/10.1007/s13280-020-01333-9
14. Pham, D.T., Castelani, M.: A comparative study of bees algorithm as a tool for function optimisation. Cogent Eng. **2**(1), 1091540 (2015)
15. Pham, D.T., Ghanbarzade, A., Koc, E., Otri, S., Rahim, S., Zaidi, M.: Bee Algorithm a Novel Approach to Function Optimisation. Technical Note Cardiff University (2005)
16. Satheesh, A., Manigandan, T.: Maintaining power system stability with facts controller using bees algorithm and NN. J. Theoret. Appl. Inf. Technol. **49**(1) (2013)
17. Seeley, T.: Honey Bee Democracy. Princeton University Press, Princeton (2010)
18. Southwick, E.E., Pimentel, D.: Energy efficiency of honey production by bees. Bioscience **31**(10), 1981 (1981)
19. Stabentheiner, A., Kovac, H.: Honeybee economics: optimization of foraging in a variable world. Scientific Reports (2016). www.nature.com/scientificreport
20. Yuce, B., Packianather, M.S., Mastrocinque, E., Pham, D.T., Lambiase, A.: Honey bees inspired optimization method; the bees algorithm. Insects **4**(4), 646–662 (2013). https://doi.org/10.3390/insects4040646
21. Carolina HoneyBee Newsletter. https://carolinahoneybees.com/the-honeybee-dances-bust-a-move/
22. Pinterest Visulisation 1. https://pl.pinterest.com/pin/331366485060584772/

23. Beguling History of Bees. https://www.scientificamerican.com/article/the-beguiling-history-of-bees-excerpt/
24. World honey production. www.fao.org/faostat
25. Honey bee smart. https://owlcation.com/stem/How-Smart-are-Honey-Bees
26. Honey bees and energy lines. https://wicklowbeesandhoney.wordpress.com/2018/08/09/bees-and-energy-lines/
27. Bees and energy lines. https://www.npr.org/2013/02/22/172611866/honey-its-electric-bees-sense-charge-on-flowers?t=1622588913001
28. Group decision making in honey bee swarms. www.americanscientist.org/article/group-decision-making-in-honey-bee-swarms\
29. Facts about honey bees. https://www.beepods.com/101-fun-bee-facts-about-bees-and-beekeeping/
30. Flower electric signals and bees. https://www.npr.org/2013/02/22/172611866/honey-its-electric-bees-sense-charge-on-flowers?t=1622588913001

The Application of Artificial Intelligence to Nuclear Power Plant Safety

Ceyhun Yavuz and Senem Şentürk Lüle$^{(\boxtimes)}$

Energy Institute Sariyer, Istanbul Technical University, 34467 Istanbul, Turkey
senturklule@itu.edu.tr

Abstract. Through application of artificial intelligence (AI), the burden of analytical computational load for analysis of any given problem where countless variables have to be taken into account, is virtually eliminated. Since for engagements in real life operations and instantaneous actions are of paramount importance and vital, AI can be a strong alternative to overcome the complex problem solving in short time frames. As such, in this study a brief review of AI basics is given and literature for AI applications in nuclear field such as defect detection in nuclear fuel assembly, dose prediction in nuclear emergencies, fuel and component failure detection, core monitoring for reactor transients, core fuel optimization models, gamma spectroscopy analysis and specifically nuclear reactor safety in operation are assessed. Afterwards, an AI model for analyzing transients in VVER type nuclear power plants that is being built in Turkey is proposed. This model must keep up with instantaneous data flow and giving actionable feedback to the operator both for the cause and the solution. A semi-autonomous AI control system that help the operator decision making is a significant contributor to the safety of a reactor.

Keywords: Artificial intelligence · Artificial neural networks · Nuclear reactor safety · Nuclear reactor operation · VVER 1200 · Transient analysis

1 Artificial Intelligence

Artificial Intelligence (AI), the art of making a computer learn like a human, is a strong alternative for addressing and analyzing complex real world problems. While Rule Based-Expert Systems are valid approaches, they do not perform well in terms of encompassing the volatile and co-dependent parameters of the physical problems. Machine Learning (ML) is a sub-branch of AI which allows code to learn through assessing data and predicting or forecasting from accumulated experience of the set. Although there are many classifications, ML has three main classes.

1.1 Supervised Learning

Supervised learning works with a labelled data set, a set of observations associated with their respective outcomes [1]. Deriving a function from training data set, model predicts an outcome which is classified as regression or predict a label/category which is classified as classification.

© IFIP International Federation for Information Processing 2022
Published by Springer Nature Switzerland AG 2022
E. Mercier-Laurent and G. Kayakutlu (Eds.): AI4KMES 2021, IFIP AICT 637, pp. 117–127, 2022.
https://doi.org/10.1007/978-3-030-96592-1_9

1.2 Unsupervised Learning

Unsupervised learning uses a set of algorithms to infer a conclusion from data sets without an outcome. Cluster analysis is the most common unsupervised learning method, which analyzes pattern and groupings in the data set [2].

1.3 Reinforced Learning

Reinforcement learning assigns a goal in a complex environment where model employs trial and error. For every result it gets a reward or a penalty, thus reaching a robust model. Designer sets the policy for reward-penalty but does not guide the model towards a specific model [2].

1.4 Artificial Neural Network

The Artificial Neural Network (ANN) is based on a collection of connected units or nodes called artificial neurons, which loosely model the neurons in a biological brain as seen in Fig. 1 [3]. Each connection, like synapses in a biological brain, can transmit a signal to other neurons. An artificial neuron receives a signal then processes it and can signal neurons connected to it. The "signal" at a connection is a real number, and the output of each neuron is computed by some non-linear function of the sum of its inputs. Neurons and edges typically have a weight that is adjusted as learning proceeds. The weight increases or decreases the strength of the signal at a connection [4]. Neurons may have a threshold such that a signal is sent only if the aggregate signal crosses that threshold. The adjective "deep" in deep learning refers to the use of multiple layers in the network [5].

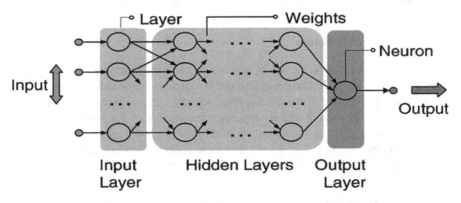

Fig. 1. Exemplary artificial neural network structure [3]

2 Applications in Nuclear Field

2.1 Defect Detection in Nuclear Fuel Assembly

Fuel assembly is made out of fuel rods where fission reaction occurs. The clad of fuel rods carries high thermal and mechanical requirements and standards, since its integrity is directly correlated with nuclear power plant (NPP) safety. There is a regular inspection program for inspection of fuel assemblies to maintain safety standards. These in service inspections are executed manually and visually (camera assisted). Since this is a time consuming and subjective procedure, it is prone to errors. An alternative is the sipping test that determines whether the fuel assembly contains a defect by checking the amount of Kr-85 radioisotope contained in the fluid sample. But the sipping test cannot be executed for in operation assemblies.

To optimize and deliver accurate results, deep convolutional neural network was proposed to detect the scratches on nuclear fuel assembly. The proposed framework achieves 0.98 TPR (True Positive Rate) against 0.1 FPR (False Positive Rate) that is significantly higher than most of current detection method. A test case for trained R-CNN in Fig. 2, demonstrates the model's accuracy, not only by distinguishing between real defects thus resulting in a high TPR, but also between water stains and other false positives thus resulting in a low FPR [6].

Fig. 2. The detected image of testing data [6]

2.2 Dose Prediction in Nuclear Emergencies

NPP accidents can result in release of radioactive materials to the atmosphere. To predict the atmospheric dispersion and equivalent doses with respect to distance from accident site is vital since evacuation of population and assessment of the environmental impact are key parameters [7]. Deriving this result analytically can be done with great computational cost.

In order to predict spatial effective doses for a wide range of atmospheric conditions, the Deep Rectifier Neural Network (DRNN) was evaluated considering two realistic accident scenarios simulated with the atmospheric dispersion system used in a Brazilian NPP. In the more simplified scenario, the best DRNN achieved an error 25% lower than the 5-MLP with a training 155 times faster. In the more complex scenario, the best DRNN achieved an average error of 0.0213 with a training time of 30 min, demonstrating that DRNNs can improve ANN-based dose prediction in realistic situations [8].

2.3 Fuel and Component Failure Detection

In NPP, fuel cladding is what keeps the radioactive materials in a closed cascade. In case of failure of this component, the release of radioactive materials to primary circuit is inevitable. The common method for failure detection is isotopic ratios method where the ratio of released isotopes to the coolant is checked for and if it is above a threshold, the detection of the failure takes place.

Four-layer ANN, which consists of an input layer, two hidden layers, and an output layer was proposed for fuel failure detection as seen in Fig. 3. The neurons receive total values from the preceding layer and transmit to the next layer by the activation function. The inputs of the ANN are normalized specific activities of 23 typical fission products, include Br-83, Kr-85, Kr-85 m, Kr-87, Kr-88, Sr-90, Te-131, Te-131 m, I-129, I-131, I-134, I-135, Xe-133, Xe-133 m, Xe-135, Xe-135 m, Xe-138, Cs-134, Cs-134 m, Cs-137, Cs-139, Pr-143 [9].

Comparing to the isotopic ratios method, the ANN proves to be more responsive when the fuel cladding is defective.

2.4 Core Monitoring for Reactor and Transients

A real-time monitor for 3D reactor power distribution is critical for nuclear safety and high efficiency of NPP's operation as well as for optimizing the control system, especially when the nuclear power plant (NPP) works at a certain power level or it works in load following operation [10]. Also NPP experiences transients which may occur be due to equipment failure or malfunctioning of process systems. The operator has to carry out diagnostic and corrective actions with respect to determined transient. ANN performs diagnosis fast and within acceptable error margin. This presents the opportunity of reduction in human evaluation error and fast response to the scenario at hand. A benchmark for identification performance of the transients together with initiating events is listed in Table 1. With these error rates, after decreasing it to an acceptable level, a decision support system becomes feasible and executable [11].

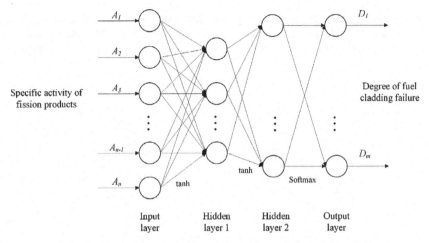

Fig. 3. The architecture of the ANN for fuel failure detection [9]

Table 1. IES and their average identification errors [11]

Initiating event	Average identification error (%)
Steam generator tube rupture	
SG1TR (steam generator)	4.5
SG2TR (steam generator)	5.2
SG3TR (steam generator)	4.9
SG4TR (steam generator)	4.6
Moderator heat exchanger tube failure	
MODHX1TR (moderator heat exchanger)	5.3
MODHX2TR (moderator heat exchanger)	5.5
Bleed cooler tube rupture	7.4
Shut down cooling heat exchanger tube fail	8.2

2.5 Core Fuel Optimization Models

In any NPP, the optimization of nuclear fuel configuration in the core is one of key performance activities [12] and an accurate prediction on the neutronic parameters of a nuclear reactor core is a major design concept for both economic and safety reasons [13]. Due to burn-up limitations, fuel loading and shuffling generally take place annually. During refueling, the highly burned fuel assemblies are discharged form the core, the rest is shuffled and fresh fuel assemblies are placed in the core. The purpose of the refueling is to maximize the effective multiplication factor k_{eff} while keeping power peaking factor P_r below acceptable limit [14].

When Particle Swarm Optimization is utilized to derive a model, it optimized P_r and k_{eff} with the achieved pattern. As a result, an increased k_{eff} by 8.79% and a reduction in by 0.047% compared to classic reference loading pattern was achieved (Fig. 4) [15].

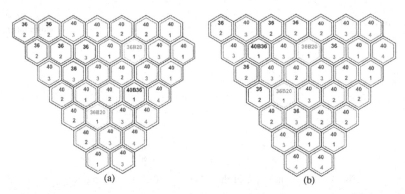

Fig. 4. Comparison of the one-sixth core loading pattern of the reference cycle, with the optimized one: (a) proposed core loading pattern, (b) reference core loading pattern [15]

2.6 Gamma Spectroscopy Analysis

Neutron activation analysis (NAA) is a method for the qualitative and quantitative determination of elements in samples based on the measurement of characteristic radiation from radionuclides formed by irradiating materials with neutrons. ANN algorithm can be usefully applied to the analysis of gamma-ray spectra by recognizing the combined pattern of emitted gamma-ray energies. In the rare situation where there is only a single gamma-ray from a radioisotope and its energy overlaps with other gamma-rays, it becomes difficult for the ANN to identify the exact radioisotope [16].

3 Artificial Intelligence Based Semi-autonomous Control System to Assist Decision Making of Reactor Operators for VVER 1200

3.1 Transients

Transient is a change in the reactor coolant system temperature, pressure, or both, attributed to a change in the reactor's power output [17]. Transients can be caused by

1. Adding or Removing Neutron Poisons
2. Change in Electrical Load on Generator
3. Accidents

An AI based control system to diagnose reactor transients and present actions to operator with respect to transient that is determined is of high value to Nuclear Reactor Safety [18]. Some transients for nuclear power plants are given below.

- **Startup and Shutdown**

In any Startup or Shutdown operation, Reactor Power, Temperature and Pressure changes.

- **Trips and Controls**

The automatic or manual shutdown of the plant is referred to as a "trip" or "Scram".

- **Starting/tripping Pump**

Pump will trip when motor is overloaded due to running of pump in runout condition. Pump may also trip due to cavitation or mechanical seal damage.

- **Loss of Coolant Accidents (LOCA)**

LOCA describes an event in which the coolant is lost from the core due to rupture of a pipe. The most dramatic scenario is known as a large-break LOCA (LBLOCA) in which a double-ended failure (often referred to as a guillotine rupture) of one of the main primary circuit coolant pipes [19].

- **Loss of Flow Accident (LOFA)**

LOFA occurs when pumping power is lost and the coolant becomes stagnant [20].

- **Loss of Offsite Power (LOOP)**

The loss of offsite power (LOOP) initiating event occurs when all electrical power to the plant from offsite power line is lost. The NPP offsite power system is the transmission power system where NPP is connected.

3.2 VVER 1200

The VVER reactor (Fig. 5) itself was developed by ROSATOM subsidiary OKB Gidropress, while the nuclear power plants employing the VVER have been developed by the power plant design organizations within ROSATOM: Moscow Atomenergoproekt. The VVER is a pressurized water reactor (PWR) with properties in Table 2, the most common type of nuclear reactor worldwide employing light water as coolant and moderator. However, there are some significant differences between the VVER and other PWR types, both in terms of design and materials used. Distinguishing features of the VVER include the following:

- Use of horizontal steam generators;
- Use of hexagonal fuel assemblies;
- Avoidance of bottom penetrations in the VVER vessel;
- Use of high-capacity pressurizers [21].

Table 2. VVER 1200 parameters [21]

VVER-1200 parameters	Values
Reactor nominal thermal power, MW	3200
Availability factor	0,9
Coolant pressure at the reactor outlet, MPa	16,2
Coolant temperature at the reactor inlet, °C	298,6
Coolant temperature at the reactor outlet, °C	329,7
Maximum linear heat rate, W/cm	420
Steam pressure at the outlet of SG steam header (abs), MPa	7,0
Primary design pressure, MPa	17,64
Secondary design pressure, MPa	8,1
Max fuel burnup fraction over FAs unloaded, MWD/kgU	70
Refueling period, month	12/(18–24)
Time of fuel residence in the core, year	4–5

Fig. 5. VVER1200 [21]

3.3 Training Data Acquisition

Production of training data is one of the main challenges for any AI system, more so for a NPP. To overcome the burden, the reactor systems will be nodalized with RELAP5 (Fig. 6) which is used to simulate the dynamics of a NPP in accidental scenarios and to generate transient datasets of reactors for training neural networks to detect severe accidents [22]. Steady state operation at full power will be simulated first to determine temperature, pressure and void fraction values of the system at critical points. Transients will then be introduced via RELAP5/MOD3 and the associated data will be stored as training data.

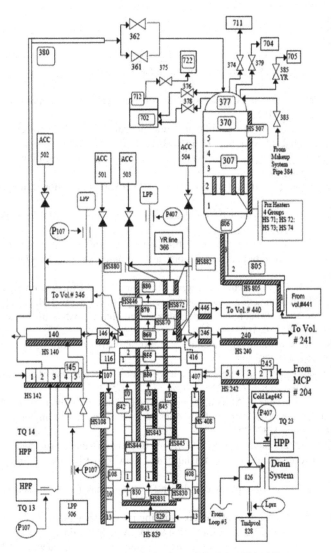

Fig. 6. Sample nodalization of VVER 1200 [23]

4 Conclusion

Artificial Intelligence (AI) is a dependable and coherent alternative to assist operation of nuclear power plants to detect transients and determine the right course of action. An AI based control system for VVER 1200 reactor would be a useful to the Turkish nuclear industry as a preliminary product since four units of VVER 1200 is being currently built in Akkuyu Site in Turkey. In addition, the Nuclear Regulatory Authority in Turkey can use this system to perform safety analysis of the reactors to control the safety analysis report calculations provided by the licensee. The system developed during this study can also be used to train future operators, free of the cost and risk associated with a real power reactor operation.

References

1. Krawczak, M.: Multilayer Neural Networks a Generalized Net Perspective, Springer Cham (2013)
2. da Silva, I.N.: Artificial Neural Networks: A Practical Course, Springer International Publishing, Cham (2017)
3. Santosh, T.V.: Application of artificial neural networks to nuclear power plant transient diagnosis. Reliab. Eng. Syst. Saf. **92**, 1468–1472 (2007)
4. Liu, W.: A survey of deep neural network architectures and their applications. Neurocomputing **234**, 11–26 (2017)
5. Abiodun, O.I.: State-of-the-art in artificial neural network applications. Heliyon **4**, e00938 (2018)
6. Guo, Z.: Defect detection of nuclear fuel assembly based on deep neural network, Ann. Nucl. Energy **137**, 107078 (2020)
7. Ling, Y.: Nuclear accident source term estimation using kernel principal component analysis, particle swarm optimization, and backpropagation neural networks. Ann. Nucl. Energy **136**, 107031 (2020)
8. Desterro, F.S.M.: Development of a deep rectifier neural network for dose prediction in nuclear emergencies with radioactive material releases. Prog. Nucl. Energy **118**, 103110 (2020)
9. Dong, B.: Detection of fuel failure in pressurized water reactor with artificial neural network, Ann. Nucl. Energy **140**, 107104 (2020)
10. Xia, H.: Research on intelligent monitor for 3D power distribution of reactor core. Ann. Nucl. Energy **73**, 446–454 (2014)
11. Saeed, A.: Development of core monitoring system for a nuclear power plant using artificial neural network technique, Ann. Nucl. Energy **144**, 107513 (2020)
12. Nissan, E.: An overview of AI methods for in-core fuel management: tools for the automatic design of nuclear reactor core configurations for fuel reload, (re)arranging new and partly spent fuel. Designs **3**, 37 (2019)
13. Pirouzmand, A.: Estimation of relative power distribution and power peaking factor in a VVER-1000 reactor core using artificial neural networks. Prog. Nucl. Energy **85**, 17–27 (2015)
14. Rose Mary, G.P.: Neural network correlation for power peak factor estimation. Ann. Nucl. Energy **33**, 594–608 (2006)
15. Babazadeh, D.: Optimization of fuel core loading pattern design in a VVER nuclear power reactors using Particle Swarm Optimization (PSO). Ann. Nucl. Energy **36**, 923–930 (2009)
16. Sahiner, H.: Gamma spectroscopy by artificial neural network coupled with MCNP. Doctoral dissertations. p. 2598 (2017)

17. U.S. Nuclear Regulatory Commission. https://www.nrc.gov/reading-rm/basic-ref/glossary/transient.html. Accessed 31 Nov 2021
18. de Oliveira, M.V.: Application of artificial intelligence techniques in modeling and control of a nuclear power plant pressurizer system. Prog. Nucl. Energy 63, 71–85 (2013)
19. Joyce, M.: Nuclear Engineering, Nuclear Safety and Regulation, Butterworth-Heinemann, New York (2018)
20. Mogahed, E.A.: Loss of Coolant Accident and Loss of Flow Accident Analysis of the Aries-at Design, Fusion Technology Institute University of Wisconsin-Madison (2010)
21. Mokhov, V.A.: Advanced Designs of VVER Reactor Plant, VVER-2010 Experience & Perspectives 01–03 November 2010, Prague Czech Republic (2010)
22. Tian, D.: A constraint-based genetic algorithm for optimizing neural network architectures for detection of loss of coolant accidents of nuclear power plants. Neurocomputing 322, 102–119 (2018)
23. Ivanov, B.: VVER-1000 Coolant Transient Benchmark. US Department of Energy, Nuclear Energy Agency Organization For Economic Co-operation and Development (2002)

Capacity to Build Artificial Intelligence Systems for Nuclear Energy Security and Sustainability: Experience of Belarus

Yuliya Pranuza[✉]

Francisk Skorina Gomel State University, Gomel, Belarus

Abstract. In recent decades the danger of man-made hazards in the nuclear field has increased dramatically. Belarusian scientists have accumulated considerable innovative potential in health care, agriculture, the creation of new life support technologies in radioactively contaminated areas. This information is required when working out catastrophe scenarios at nuclear power plants. The definition of situational innovations is developed that are intended for use in certain circumstances and are essential for the achievement of unique goals including ensuring people's life and safety. The highlighting situational innovations makes it possible to determine their adaptation, use and replication in other economies, to diffuse them, to increase effectiveness of the cost of their working out. The results of the study indicate that (1) the unique experience of Belarus in overcoming the consequences of a nuclear disaster is of great practical importance for other countries operating nuclear objects, (2) the situational innovations resulting from the overcoming of nuclear disaster need to be consolidated into a data base and incorporated into the international data base of the results of nuclear research, (3) the application of Artificial Intelligence (AI) systems allows to accumulate, store, speed up the search of information in a single source and provide the necessary situational innovations in case of need (nuclear accident). AI systems also help to prevent and reduce the risk of nuclear accidents. Development and using AI will allow to help countries worldwide develop nuclear disaster information management systems and reduce existing disaster risk for sustainable development in the future.

Keywords: Nuclear disasters · Situational innovations · Data base · Artificial Intelligence · Knowledge management

1 Introduction

The economic success of the developed countries of the world is determined by innovation but innovation is not always good.

Human activities have become a real factor of the global negative impact on nature. Environmental pollution has become a worldwide damage. Vast territories, oceans and people suffer from the harmful emissions of industry. Different kinds of animals, insects, plants are disappearing. Often these processes become irreversible.

E. Mercier-Laurent and G. Kayakutlu (Eds.): AI4KMES 2021, IFIP AICT 637, pp. 128–140, 2022.
https://doi.org/10.1007/978-3-030-96592-1_10

Innovation creates new risks and may increase the probability of accidents, catastrophes, and poorly predicted events. Catastrophes at nuclear plants are a good example of it. Nuclear disasters have a particular negative impact on the sustainability of the ecological ecosystem.

Unfortunately, the world knows examples of such major accidents which have caused great environmental damages and are an obstacle for sustainable development: Three Mile Island (1979), Chernobyl (1986), Fukushima Daiichi (2011). These accidents have caused great environmental damages and are the obstacle for sustainable development. 'These serious accidents produce human, material losses and pollution that cannot be recovered' [1].

The aim of this article is to justify the necessity to use the possibilities of AI systems for the distributing in other economic systems the unique experience of Belarus in obtaining situational innovations that are of great importance for people's health worldwide.

As a result of the disaster at the Chernobyl nuclear power plant the economy of Belarus has suffered in considerable damages. Belarus was exposed to significant radioactive contamination after the disaster. Today, 35 years after the accident, one of the reasons of the low level of social economic development is the consequences of the nuclear disaster.

Many problems had to be solved for the first time and in a very short period of time under conditions of high radiation. Due to specific limited demand, a large number of innovations have emerged. This work is still under way. It was suggested that such innovations be allocated as a separate group-situational innovations.

Situational innovations are innovations that have been applied under certain unique circumstances that are needed in the course of the implementation of poorly studied events, and that have the potential of using in other socio-economical systems [2].

Innovations generated within certain industries (especially those with the potential for adverse events) should be available for application in relevant organizations around the world.

The need for such information is demonstrated by the considerable increase in mutual contacts between Belarusian and Japanese scientists working on the problem of reducing the effects of radiation by the results of the large disaster at Fukushima Daiichi (2011).

The analysis of feedback from such catastrophes and incidents helps improving design, maintenance and training, planet protection, security and decision about the next purchase, but cannot repair damages. It is insufficient for preventing the next disaster, because the knowledge of context is incomplete [1]. But the accumulated information certainly helps to reduce uncertainty and risk of repetition of negative events in the future in the field of nuclear plants.

The present article tries to clarify possibilities of AI systems to create, store, search for and ultimately reduce the existing nuclear disaster risk for sustainable development.

2   Experience of Belarus

The territory of Belarus was exposed to significant radioactive contamination after the Chernobyl (Ukraine) nuclear power plant disaster. The release of a large number of radionuclides has altered the natural environment in large areas.

The disaster caused the death and illness of many people, irradiation of various severity and the contamination of vast territories of Belarus, Ukraine, Russia and other countries.

As a result of the initial explosion at the Chernobyl nuclear power plant (1986) and subsequent fire after it the radionuclides were transferred from Chernobyl in Ukraine across the border to Belarus what resulted in the contamination of 47,600 km^2 (23% of the territory of the country on which 20% of its population lived [3]. Some 21% of the country's agricultural land, 23% of the forested land and 132 mineral deposits were polluted [4].

The negative impact for infrastructure and the relocation of people from polluted regions have became the reason of decrease of economic activity. The tasks of minimizing the consequences, as well as creating a safe living environment after the disaster, could not be solved on the basis of the available possibilities at that time. Innovative solutions were needed.

Specific problems that have arisen since the Chernobyl disaster (1986):

i)   the high risk of the consequences of the catastrophe especially for the popula-tion and the need to take unique decisions in a short period of time,
ii)   difficulties in assessing the effects of radionuclides on humans and the natu-ral environment,
iii)   lack of reliable means of identification, measurement and evaluation of ad-verse effects and radiation,
iv)   lack of knowledge and practice of minimizing the consequences of a major nuclear accident,
v)   application of unique technical solutions adopted in the course of post-disaster operations and constantly adaptable to specific situations,
vi)   training and professional retraining of a large number of specialists in the ap-plication of various types of technologies not previously used or described in the literature,
vii)   high level of fatal risks for the participants in the disaster,
viii)   some erroneous management decisions based on ignorance of the conditions of the situation, etc.

The practice of unprecedented and effective interaction of the main components of the innovation process (scientists and practitioners introducing new ideas have emerged). The Chernobyl problems were new for that time and required a change in scientific thinking and action even for the most authoritative scientists.

Much has been done by trial and error and usually without experimental checkup. Decisions have been taken in high risk situations. Perhaps for the first time large-scale innovations were introduced in a highly uncertain and dynamic environment.

In Belarus, specialized research institutes have been established in a short period of time. The medical centers have been built to treat the affected inhabitants from radiation exposure. The recommendations have been prepared on the maintaining of production and the prevention of diseases, radiation monitoring, etc. Wherein, the rules of living have changed. Large areas have been made uninhabitable and potentially dangerous for long-term visits.

New habitation patterns have emerged: limited use of local raw materials; the need for a reprofiling of production; changing demographics, etc.

For the territories affected by the disaster, there were specific needs that were not present on other territories, namely:

i)   radiation-safe living conditions,
ii)   'pure' food,
iii)   special medical care,
iv)   improvement of the 'clean' areas with regard to radiation,
v)   compensation for damage and risk of living in contaminated areas,
vi)   special information services,
vii)   relocation to 'clean' land,
viii) additional measures to support the development of the economy, etc.

In the Belarusian part of the exclusion of the Chernobyl zone nuclear power plant on the territory of Gomel region (Braginsky, Narovlyansky and Khoiniki region) the Polesie State Radiation and Ecological Reserve was organized (1988) [5]. The population living on this territory was evacuated and the land was taken out of economic use. The total area of the reserve is over 217,000 hectares.

The reserve was established with the aim 'of implementing measures to prevent the removal of radionuclides outside its territory, to study the state of natural vegetation complexes, fauna, to conduct radiation-ecological monitoring, and to conduct radiobiological research' [5].

The Reserve monitors and controls the status of radionuclide contaminated territory. The processes of accumulation and migration of radionuclides (^{137}Cs, ^{90}Sr, ^{241}Am) in components of terrestrial and aquatic ecosystems are studied. The forecast of radiational situation in the closest to the Chernobyl nuclear power plant zone is being projected for a more remote period.

The reserve is the experimental base where research on agricultural activity in the conditions of radioactively contaminated areas is being carried out.

Scientific work in the reserve is important for biodiversity conservation. Fauna is developed in the absence of an anthropogenic pressure. For example, the acquisition of the reserve by the bison takes place naturally without human intervention. The permanent inhabitants of the reserve are bears, deer, wolves, lynxes.

The state of vegetation communities affected by radioactive contamination is being studied in the reserve. Factors influencing radionuclide accumulation patterns in different plant species are analyzed (based on plant characteristics and soil moisture availability).

The investigation of the impact of cesium and strontium contamination on plants in the environment is carried out. Within the framework of this work the possibility of obtaining pure production of beekeeping with the soil contamination by ^{137}Cs and ^{90}Sr has been confirmed. Plants that contribute to the production of low-radionuclide honey were identified.

Consequently, Polesie State Radiation and Ecological Reserve is a unique research object that conducts research and innovates numerous safety measures in a radionuclide contaminated area.

There are also other scientific organizations based in Belarus that have important scientific results in the nuclear field.

Such organizations include Institute of Radiobiology of the National Academy of Sciences of Belarus (Gomel), National Centre for Radiation Medicine and Man's Ecology (Gomel), Scientific and Practical Center of the National Academy of Sciences of Belarus for Bioresources (Minsk), etc. [6–8].

For example, it was developed the contamination prediction model of cesium-137 and strontium-90 concentrations in crop production in the long term after Chernobyl nuclear accident [9]. The soil-to-plant transfer of radionuclides strongly depends on agrochemical characteristics of soils. The model offers a set of changeable soil parameters (acidity, humus content, exchangeable potassium and phosphorus) to make a more accurate prediction of ^{137}Cs and ^{90}Sr contamination for a selected variety. This prediction model allows to predict the accumulation the ^{137}Cs and ^{90}Sr in agricultural crops grown on contaminated soils of different types (on soddy-podzolic and peaty soils) in the long term after the Chernobyl accident.

It was developed the web application specifically designed for the calculation of prediction values of ^{137}Cs and ^{90}Sr concentrations for a wide range of crops cultivated on soddy-podzolic and peaty soils. The prediction model app is available in Russian only (http://forecastmodel.pythonanywhere.com/).

It was developed the agricultural recommendations for radiation-affected farmlands by the belarusian scientist. For example, the Recommendations on the application of mineral fertilizers for the perennial mid-season grass mixture cultivation on ^{137}Cs-contaminated peaty soils. And, the Recommendations for home crop cultivation and personal livestock farming in contaminated rural areas in the long term after the Chernobyl accident [10].

The innovations were created in Belarus which are needed in the situations of manage the nuclear accident and prediction the development the situation. These innovations have been tested in real situations of radiation.

The results of the research carried out by belarusian scientist and organizations should be applicable in other countries as they relate to life and health.

Consequently, Belarus had acquired unique experience and proposed innovative way of managing disasters. These innovations are extremely rare and are essential for achieving unique goals, including the lives and safety of people, the environment and animals.

Most importantly, these innovations are of important for people's lives and health. These innovations should be considered as situational innovations.

3   The Situational Innovations

Situational innovations are driven by a specific limited demand in the conditions of unique challenges.

The situational innovations have the potential for further application in other contexts and socio-economic systems [2]. Wherein situation is 'all the circumstances and things that are happening at a particular time and in a particular place' [11].

The situational innovations in the result are unique. Uniqueness can be defined as follows: 'Being the only one of its kind; unlike anything else' [12].

Situational innovations are not applied under normal conditions. There should be certain characteristics that are appropriate to the types of situational innovation that justify their use and their adaptation (if necessary). These include natural disasters (earthquakes, floods, tsunamis, pandemics), space and Arctic exploration, and other unique projects (the Large Hadron Collider).

The following features of situational innovation have been identified:

i)   the need for working out a response for solving unique problems in the absence of knowledge,
ii) relatively low demand under special conditions (catastrophe),
iii) the need to solve the problems in the shortest possible time,
iv) the possibility to adapt and reserve further use in other economic systems and situations,
v)   financing, as a rule, with the participation of the state.

Situational innovations can be applied to disaster management in other countries, both commercially and non-profitably.

In the first case, Belarus can commercialize the results of its research in this areas.

In the second case, the results can be provided free of charge (as in the case after the accident on Fukushima Daiichi when scientific contacts between Belarusian and Japanese scientists have increased significantly).

Situational innovation can be commercialized when linked to tools of disaster management.

At the same time, in case of increasing importance of such knowledge (elimination of current accidents) situational innovation can be provided free of charge.

There is a scientific interest in classification of situational innovations due to their importance.

Situational innovations can be used within one organization or a group of organizations, only in certain industries or regions, in certain countries, or be in common use in organizations around the world (based on their possible scale of use). This approach highlights the innovations of level 1, 2, 3 and 4 (Table 1).

As an example of level 1 innovation (Table 1) the Large Hadron Collider should be called which is the largest and only experimental facility in the world to study the results of the acceleration of protons, heavy ions (lead ions) and the products of their co-bumping. This construction is unique in its kind so the innovations of the scientific experiment are assigned to level 1 innovation (Table 1).

Specific innovations applied to specific climatic conditions and only by selected countries have been developed for the development of the Arctic (level 2 Table 1). Competition for the wealth of the Arctic is tough enough, and the necessary information can be only provided on a cost-recovery basis.

It is generally accepted that the most dangerous objects created by mankind are atomic objects.

Unfortunately, the practice proved that they are not completely safe. There have been major nuclear reactor accidents around the world.

Table 1. Classification of situational innovations according to possible scale of use (unique situation)

Level of using innovations	Scale of using innovations	Examples
Level 1	In single organizations	Innovations are required in one or more organizations (use of the Large Hadron Collider and thermonuclear fusion plants)
Level 2	In selected countries	Arctic development, space exploration
Level 3	Only in a single industry/region	Nuclear energy, space exploration, radiation pollution and poisoning of territories
Level 4	In demand worldwide	Emergency and other monitoring systems, pandemic (Covid vaccination)

Therefore, the innovations generated in certain industries (especially those with the potential for negative phenomena) should be able to be applied in relevant organizations around the world (level 3 Table 1). If the Japanese Government had had the information concerning the Chernobyl accident study the consequences of the technological disaster of Fukushima Daiichi would have been lower.

The global response to atomic catastrophes has generated a significant amount of innovations which are now dispersed in different countries and organizations. It is suggested to form a data base of situational innovations obtained in the process of the implementation of measures to minimize the consequences of atomic catastrophes on the basis of an international organization (United Nations Office for Disaster Risk Reduction).

Level 4 innovations (Table 1) include innovations used in many countries around the world. These can be technological (product and process), organizational, and environmental innovations that, because of their specifies, can be applied by a wide range of organizations. Examples include an emergency monitoring system (floods, earthquakes) and a vaccine that every person on the planet is entitled to receive during a pandemic.

Situational innovations have the potential to be adapted and applied in a similar way to the initial unforeseen (unplanned) situation that gave rise to them. Accordingly, it is advisable for certain unique situations to draw up a list of characteristics that define and characterize specific situations.

Based on the results of a specific unique situation, it is necessary to establish a list of innovations, the use and adaptation of which will be appropriate in future.

This can be demonstrated by the Chernobyl disaster.

Table 2 contains a list of characteristics that define the situation at the beginning of its implementation.

It also identifies areas of innovation which can be adapted and applied to a situation with similar basic determinants. Other situational innovations may also receive similar systematization.

Table 2. List of situational characteristics and directions of innovation (situation is the disaster at a nuclear power plant)

List of characteristics	Directions of innovation
Changing environmental characteristics (radiation pollution)	Medical aspects of life in the territory of radioactive contamination
Human radiation exposure	Production of suitable agricultural products (protective measures)
Need for special protective measures	Animal husbandry in the face of radioactive contamination
Biomedical restrictions on economic activities	Activities in exclusion and eviction zones (decontamination works)
Evacuation of the population from the disaster epicentre	Ensuring safe living conditions in areas of radioactive contamination

4 Artificial Intelligence Systems for Nuclear Energy Security

Nuclear disaster risk can be facilitated by the use of AI systems. These systems make it possible to control and managed nuclear power plants safety, improve forecasting and establishing support system for radiation accidents, create nuclear knowledge management system, develop science, create robotics technology. Consider the directions of the AI in detail.

4.1 Safety Control and Managing Nuclear Power Plants

AI systems can be applied for nuclear power plants surveillance and diagnosis of problems on nuclear power plants.

AI allows to perform fast analysis of data coming from special cameras in real time based on the technology of computer vision and analysis of video images AI permits to compare them with reference data of the information database and to determine the presence or absence of a problem (for example, defects in reactor).

AI also calls for the creation of digital duplicate of nuclear power stations. Such 'digital doppelgangers' accumulate data from sensors installed on a nuclear reactor and allow to create digital copies of the entire nuclear device process.

Consequently, AI systems is a basic technology for controlling safety nuclear power plants which minimizes human factors and ultimately improves the safety and security of the nuclear power plant.

4.2 Improving the Forecasting and Decision-Making Support System for Radiation Accidents

The AI allows considering a large number of factors that have preceded or accompanied accidents in the past.

AI allows to analyze disaster impacts taking into account a large range of socio-economical aspects based on the analysis of this information and the results of monitoring of current situation.

AI provides an opportunity to simulate the spread of the radioactive cloud after the accident at the nuclear power plant, to develop an algorithm of accident prevention, to predict risk of disaster and to offer early warning.

AI can be used to improve existing decision-making support system for radiation accidents. For example, such as the system RODOS (Real time On-line Decision Support System). This system is applied for off-site emergency management after nuclear accidents for more than ten years in European and Asian countries [13].

4.3   Nuclear Knowledge Management

The nuclear industry is knowledge-based and depends on the exchange of information and experience in the design, construction, operation and decommissioning of nuclear objects.

AI is a tool for managing a great amount of nuclear data, information and knowledge.

AI systems improve the organization of data and information by linking different sources together. In this case they can be mutually shared and reused in different countries among many organizations of different types.

The results of previous research may become more accessible to stakeholders in different countries and are based on the development of distributed knowledge bases. This is really important in the context of Open Science.

Over the past decades experiences have demonstrated that knowledge management powered with AI bring considerable help in decision making related to risks prevention and management in the nuclear areas. It is vital to prevent efficiently such accidents using all available information.

To avoid human and others loss it becomes necessary to learn how to recognise and evaluate the early warning signs, gather all available knowledge in aim to reduce loss and prevent risks (when it is possible).

As an example, the International Nuclear Information System (INIS) can be given [14]. INIS is a unique information resource that provides access to nuclear information from all around the world. INIS maintains one of the world's largest collections of publications on peaceful use of nuclear science.

4.4   Development of Science

AI is dedicated to improving the system of Nuclear Knowledge Management. The use of AI system is important for the processing data of scientific experiments taking into consideration numerous data already accumulated in the past.

AI can also be applied to calculation of physical processes and can accelerate calculation of changes in physical performance of a reactor.

The AI system brings together knowledge, helps make new discoveries and reduce the overall cost of R&D.

Nuclear safety research and development will become easier and faster and may lead to new scientific breakthrough and new knowledge in this field.

4.5 Robotics Technology

AI enables to develop robotic systems which explore and manipulate objects in extreme environment such as nuclear industry.

Robotics technology can be used in the entire nuclear cycle: operating reactors, building new reactors, decommissioning and waste storage.

The robots may be involved in dangerous or inaccessible locations (for example, dismantling reactors of failed nuclear plants or working with radioactive substances).

Cleaning up radioactive waste is a dangerous job for humans. In this case, robots for nuclear power plants are just irreplaceable.

The robots make it possible to study the level of radiation in places where the molten fuel is located.

Developing robots for solving critical nuclear issues is being actively developed nowadays.

For example, there is Department of Electrical and Electronic Engineering in the University of Manchester. The laboratory of Robotics for extreme environments has developed robots for nuclear power stations adapted to the conditions of the nuclear power plant in Japan [15].

AI can find other applications in nuclear power that we can only have a guess about today.

Finally, AI systems can take over the functions of station operator and all that is left for a man is the role of observer. Eventually, this will improve the reliability and safety of operating nuclear power plants.

5   Conclusion

Belarus was exposed to radioactive contamination after the Chernobyl disaster. Many problems had to be solved for the first time and in an extremely short period of time in conditions of high radiation.

Thanks to specific limited demand, a large number of innovations have emerged. Thus, a group of innovations was created in Belarus which are urgently needed in specific situations. Many of these innovations have been tested in real situations of radiation contamination and are of global importance.

The article justifies the necessity of allocating situational innovations into a separate classification group.

The use of situational innovations depends on the purpose, the critical period of application, the need to adapt to new developments, the need for special scientific support, limited use, lack of precedents in specific cases, etc.

The allocation of situational innovations allows to determine the directions of their adaptation and replication in other economic systems, ensure their diffusion, solve unique problems in a short period of time.

Belarus has accumulated unique experience in dealing with the severe consequences of the nuclear catastrophe. The acquired knowledge and experience are relevant and can be replicated in other regions.

The relevance of the suggested proposal was also confirmed by international experts. For example, in the OECD's review of Belarus «Innovations for Sustainable Development» it is noted: 'Domestic companies have acquired unique experience as, for example, in measuring radiation after the Chernobyl accident' [3].

Limited fuel and energy resources as well as the extremely high energy intensity of the economy were important prerequisites for the construction of a nuclear power plant on the territory of Belarus. The operation of any nuclear power plant has the potential of having an impact on the environment. In connection with the introduction of the Belarusian nuclear power plant (2020), there is an awareness of the need to expand radiation monitoring, contingency plans and the implementation of all recommendations of the International Atomic Energy Agency (IAEA) to ensure the necessary safety measures.

By creating a digital data base based on AI, it is possible to provide access to such innovations for international organizations. This will make it possible to quickly obtain the necessary information from a single source in case of need (especially in critical situations).

The innovations were created in Belarus can be apply in international cooperation of nuclear power. For example, The World Association of Nuclear Operators (WANO) is a particularly means of international assistance. These assistances include establishment of a worldwide integrated event response strategy. WANO membership involves all of the world's nuclear power plant operators and other organisations involved with nuclear safety. WANO's peer reviews represented a shift from just accident prevention to mitigation [16]. The IAEA Global Nuclear Safety and Security Network is allowing its members to share nuclear safety and security knowledge and services to further the goal of achieving worldwide implementation of nuclear safety and security [17].

The availability of information in case of nuclear disasters will help prepare for new emerging hazards and risks.

The proposal is of important practical significance for sustainability around the world taking into account current trends in the construction of small nuclear power plants and the production of small nuclear weapons.

The article also explains the need to use AI systems for managing nuclear power plants.

AI enables storing information about all processes of a nuclear plant and control them in real time, ensuring safe operation of a nuclear power plant.

The prevention and management of risks of a nuclear power plant require understanding of all facets and the choice of right decision.

Prediction using AI enables anticipation of potential problems and control of risks. It is necessary unpredicted risks such as nuclear plant accident is managed.

Decision making process in the nuclear field has to consider all risks that apply in a given situation (especially in critical situations).

AI will facilitate their collection for preservation, sharing, and reusing for a variety of applications.

'While there should not be elimination of human from the decision making process. Several accidents and disasters with fully automated systems demonstrated that the cognitive capacity of humans is necessary for now' [1]. In this situation AI is needed. The final decision should do human.

We have necessity of implementing the synergy of combining of human and computer capacities (AI) from the decision making process in the nuclear sphere.

Using AI the nuclear areas может быть положено в развитие Early Warning System Concept [18].

This concept facilitates the use of scenarios to show how and by which methods unforeseen and undesirable developments can be identified early. This approach allows the understanding of the likely origins of the undesirable developments, allowing rapid intervention.

Early warning and planning systems are an important for the early detection of undesirable developments or sudden and devastating events.

'Regardless of the area of use, an early warning and planning system is generally a combination of four elements: human intelligence, supporting analytical tools, planning scenarios and a monitoring mechanism clean' [18].

Using AI will allow to help countries worldwide develop nuclear disaster information management systems and reduce existing disaster risks for sustainable development in the future.

Acknowledgements. This work was supported by the Department of Economics and Management of the Francisk Skorina Gomel State University in Belarus (Gomel).

I would like to give my thanks to my scientific mentor Valeriy Sialitski who is a well-known public figure in Gomel region and a former chairman of the Gomel Regional Council of Deputies for the opportunity for this scientific work to be completed.

I would like to thank Dmitry Shamrov, the Director of the Centre for Science, Technology and Business Information for consideration of a proposal to create a date base of situational innovations that are obtained in Gomel region based on the results of scientific research on overcoming the consequences of the nuclear disaster.

I would also like to thank Madam Eunika Mercier-Laurent for inviting me to this conference.

References

1. Mercier-Laurent, E., Rousseaux, F., Haddad, R.: Preventing and facing new crisis and risks in complex environments. Int. J. Manag. Decis. Making **17**(2), 148–170 (2018)

2. Pranuza, Y., Sialitski, V.: Situational innovation: features and significance. Sci. Innov. **4**(194), 29–32 (2019). https://doi.org/10.29235/1818-9857-2019-4-29-32

3. United Nations Economic Commission for Europe: Innovation for Sustainable Development Review of Belarus (2017). https://unece.org/economic-cooperation-and-integration/publicati ons/innovation-sustainable-development-review-belarus. Accessed 07 Apr 2021

4. United Nations Economic Commission for Europe: 3rd Environmental Performance Review of Belarus (2016). https://unece.org/ru/environment-policy/publications/3rd-environmental-performance-review-belarus. Accessed 04 July 2021

5. Polesie State Radiation and Ecological Reserve. https://zapovednik.by/#. Accessed 21 Apr 2021

6. Institute of Radiobiology of the National Academy of Sciences of Belarus. https://www.irb.basnet.by/en. Accessed 23 Apr 2021

7. Republican Research Center for Radiation Medicine and Human Ecology. http://www.rcrm.by/eng/index.html. Accessed 27 Apr 2021

8. Scientific and Practical Centre of the National Academy of Sciences of Belarus for Bioresources. http://biobel.by. Accessed 15 Apr 2021
9. Institute of Radiobiology of the National Academy of Sciences of Belarus: The Contamination Prediction Model. https://www.irb.basnet.by/en/developed-products/information-applic ations-and-materials/a-contamination-prediction-model-of-cesium-137-and-strontium-90-concentrations-in-crop-production-in-the-long-term-after-the-chernobyl-nuclear-accident/. Accessed 16 May 2021
10. Institute of Radiobiology of the National Academy of Sciences of Belarus. https://www.irb. basnet.by/en/developed-products/recommendations/. Accessed 22 May 2021
11. Situation: The OxfordDictionaries.com (2021). https://en.oxforddictionaries.com/definition/ Situation. Accessed 05 May 2021
12. Unique: The Oxford English Dictionary. https://www.lexico.com/definition/unique. Accessed 07 Aug 2021
13. Real time On-line Decision Operation System. https://resy5.iket.kit.edu/RODOS. Accessed 22 Apr 2021
14. International Nuclear Information System. https://inis.iaea.org/search. Accessed 06 May 2021
15. Department of Electrical and Electronic Engineering of The University of Manchester. https:// www.eee.manchester.ac.uk/research/themes/robotics-for-extreme-environments. Accessed 17 Apr 2021
16. World Association of Nuclear Operators. https://world-nuclear.org/information-library/cur rent-and-future-generation/cooperation-in-nuclear-power.aspx. Accessed 22 May 2021
17. Global Nuclear Safety and Security Network. https://www.iaea.org/services/networks/global-nuclear-safety-and-security-network. Accessed 27 May 2021
18. United Nations Economic Commission for Europe: Pathways to Sustainable Energy – Early Warning System Concept (2019). https://unece.org/DAM/energy/se/pdfs/Pathways_to_SE/ Early_Warning_System_Concept.pdf. Accessed 08 Aug 2021

Automated Planning to Evolve Smart Grids with Renewable Energies

Sandra Castellanos-Paez[1,2]([✉]) [ID], Marie-Cecile Alvarez-Herault[2],
and Philippe Lalanda[1]

[1] Univ. Grenoble Alpes, CNRS, Grenoble INP (Institute of Engineering Univ.
Grenoble Alpes), LIG, 38000 Grenoble, France
`sandra.castellanos@univ-grenoble-alpes.fr`
[2] Univ. Grenoble Alpes, CNRS, Grenoble INP, G2Elab, 38000 Grenoble, France

Abstract. Smart electrical grids play a major role in energy transition
but raise important software problems. Some of them can be efficiently
solved by AI techniques. In particular, the increasing use of distributed
generation based on renewable energies (wind, photovoltaic, among others) leads to the issue of its integration into the distribution network.
The distribution network was not originally designed to accommodate
generation units but to carry electricity from the distribution network
to medium and low voltage consumers. Some methods have been used
to automatically build target architectures to be reached within a given
time horizon (of several decades) capable of accommodating a massive
insertion of distributed generation while guaranteeing some technical
constraints. However, these target networks may be quite different from
the existing ones and therefore a direct mutation of the network would
be too costly. It is therefore necessary to define the succession of works
year after year to reach the target. We addressed it by translating it to
an Automated Planning problem. We defined a transformation of the
distribution network knowledge into a PDDL representation. The modelled domain representation was fed to a planner to obtain the set of
lines to be built and deconstructed until the target is reached. Experimental analysis, on several networks at different scales, demonstrated
the applicability of the approach and the reduction in reliance on expert
knowledge. The objective of further work is to mutate an initial network
towards a target network while minimizing the total cost and respecting
technical constraints.

Keywords: Automated planning · Smart grids · Distribution
network · Distributed generation

1 Introduction

Distribution power grids have historically been developed to deliver electricity
from the transmission grid to the final customers. For instance, in France, the
total length of the distribution grid is 1,377,269 km against 105,942 km for the

© IFIP International Federation for Information Processing 2022
Published by Springer Nature Switzerland AG 2022
E. Mercier-Laurent and G. Kayakutlu (Eds.): AI4KMES 2021, IFIP AICT 637, pp. 141–155, 2022.
https://doi.org/10.1007/978-3-030-96592-1_11

transmission grid i.e. the slightest investment must be justified by a technical and economic calculation. In a context of unidirectional power flows with customers having a relatively well-known electrical behaviour, rules for the construction and development of the grid have been defined based on realistic assumptions of load growth.

Many changes have started to appear in recent years at varying speeds depending on the country. First of all, distributed generations (DGs) based on renewable energies have increased considerably due to various regulations aiming at increasing their penetration rate. Because of their small size, these productions are connected to the distribution grid which was not initially sized to accommodate a high amount of DGs. In addition, the decarbonization of transport is leading to the development of electric vehicles, which will form a new significant load on the distribution grid that could increase peak consumption significantly without a smart operation. These changes could cause over-voltage or over-current constraints in the distribution grids, requiring significant investments in order to strengthen the grid to make it more robust.

In order to anticipate these major changes, distribution grid operators (DSOs) aim to define the optimal long-term grid architecture (in a horizon of a few decades) called target grid, optimising a set of technical and economic performance indicators while respecting a set of constraints they have predefined. These modifications can be *minor*, by adding and/or removing lines and transformers, *intermediate* by creating new parts connected to the existing grid (expansion planning) or *major*, by changing the complete architecture (greenfield planning) [9,14]. Once the target grid has been determined, the set of intermediate grids allowing the transition from the initial grid to the target grid has to be defined.

Even if decision-making support tools for DSOs exist, such as calculation modules allowing the evaluation of different performance indicators, there is no tool allowing to automate the creation of target grids as well as the succession of intermediate grids. We believe that AI Planning can be applied to determine these intermediate grids. Indeed, AI planning is beneficial to address problems subject to continuous change, a large number of high-performance planners (implementing different approaches) are already available and planning algorithms are constantly evolving.

In this paper, we investigate how AI planning techniques can be leveraged to address the evolution of smart grids. More specifically, we address the dynamics of line connections of a distributed network. This task was previously tackled by hand and therefore with this work we enable new levels of automation. More precisely, in Sect. 2 we present the context of distribution systems while in Sect. 3 we introduce the AI planning notions used in this work. In Sect. 4, we show our initial investigation on how AI Planning can be used for Smart Grids. Finally, we conclude in Sect. 5 by discussing our results and giving some perspectives.

2   Distribution System Planning Rules

In France, the most common traditional architecture is the *secured feeder* shown in Fig. 1. Each primary substation (square) is connected to another primary substation via a set of electric lines, called main feeders (lines), supplying secondary substations (small and big circles). Two connection modes are possible depending on whether the grid is in a rural or urban environment. In rural areas, the load density being low, the main feeder passes close to the secondary substations which are connected to them via secondary feeders also called antennas. In urban areas, the load density being high, the main feeder is directly connected to secondary substations. Each secondary substation is connected to the main or secondary feeder via two remote-controlled or manual switches. For the grid to be radial, all of these switches are normally closed (small circle) except one (in the big circle). The advantage of having a loopable grid is to be able to reconfigure it in case of fault. Normally *open* switches are always remotely controlled, but for normally *closed* switches this is only the case for a few (techno-economic compromise). Switches are manually or remotely operated in order to isolate the faulty portion of the feeder so that customers can be re-energised while field crew carry out repairs. In this work, we consider grids in urban environments. We also assume that the provided target grid, optimising a set of technical and economic performance indicators while respecting a set of constraints, is reachable from the provided initial grid. In other words, both grids have the same primary and secondary substations and differ only by their connections and their normally open switches. Additionally, we are focused in the transition from the initial grid to the target grid, i.e. the intermediate grids. Finally, in this initial investigation, we only address the dynamics of line connections.

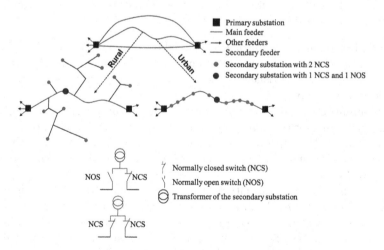

Fig. 1. Secured feeder architecture.

3 Automated Planning

Automated planning (AI planning), a sub-field of Artificial Intelligence, is a model-based approach to action selection [6]. It aims to study and design domain-independent general approaches to planning[1] [7]. The dynamics of the domain of interest, their possible actions and the conditions to attain some goal are expressed on a high-level description of the world, namely *the planning model*. By using AI planning, the development of a domain-specific solver, which would cost time and money, is not necessary – it comes down to write the planning model. Thus, AI planning represents a cost-effective method for quickly setting up a solver since planning models can be more human readable and easier to modify. In the following, we present the formal definition of the planning key concepts used in this work.

3.1 Key Concepts

Because the interest of planning lies in choosing actions to transform the system state, the transitions between states are represented with a state-transition system model. A state s is a set of predicates, i.e. a set of logical propositions. We address sequential planning in the STRIPS framework [5].

A planning task consists of a planning domain Σ and a planning problem $\mathcal{P}$. A classical planning domain is a restricted state-transition system $\Sigma = (S, A, \gamma)$ such that:

- S is included in the set of all states that can be described with the representation language $\mathcal{L}$.
- A is the set of all actions a.
- $\gamma(s, a)$ is the state-transition function that defines the transition from a state s to an state s' using an action a.

A classical planning problem $\mathcal{P}$ is defined over a domain Σ as $\mathcal{P} = (\Sigma, s_0, g)$ being s_0 an initial state where $s_0 \in S$ and g a goal, namely a set of instantiated predicates. A goal is satisfied if the system attains a state s_g such that all predicates in g are in s_g.

A planning operator is a triple $o =$ *(name_o, precond_o, effects_o)* where *name_o* is in the form *name_o*$(x_1, ..., x_n)$ such that $x_1, ..., x_n$ are the object variable symbols that appear in o, *precond_o* is the set of predicates that must hold before exploiting the action and *effects_o* is the set of predicates to be applied to a state.

An action a is an instantiation of a planning operator. Thus, a is a triple $a = (pre\_a, add\_a, del\_a)$. If an action can be applied, a new state is generated. First it deletes all instantiated predicates given in the delete list del_a, also known as the negative effects. Then, it adds all instantiated predicates given in the add-list add_a, also known as the positive effects.

---

[1] Here we refer to planning as the problem of finding a sequence of actions to achieve a goal.

A state s' is reached from s by applying an action a according to the transition function in (1).

$$s' = \gamma(s, a) = (s - del(a)) \cup add(a). \tag{1}$$

The application of a sequence of actions $\pi = \langle a_1, \ldots, a_n \rangle$ to a state s is recursively defined in (2).

$$\gamma(s, \langle a_1, \ldots, a_n \rangle) = \gamma(\gamma(s, a_1), \langle a_2, \ldots, a_n \rangle). \tag{2}$$

A *satisfying* plan is an ordered sequence of actions $\pi = \langle a_1, \ldots, a_n \rangle$ such that $s_g = \gamma(s_i, \pi)$ satisfies the goal g and the latter is reachable if such a plan exists. The plan is *optimal* if it is the shortest possible path.

3.2 PDDL Representation Language

The planning model is written using a representation language. One of the languages used in AI Planning (and selected for this work) is called PDDL. It stands for Planning Domain Definition Language [12,13]. It was introduced in 1998 for the International Planning Competition with the aim of standardising the planning representation language.

The planning model is then written as a compact representation of a planning task in PDDL-Code. The domain is composed of *predicates* which characterise the properties of the objects and a set of non-instantiated *actions* which establish the ways to move from one state to another. The problem is composed of *objects* which define the task relevant things in the world; an initial state s_i which represents the starting configuration of the world; and a goal state g which describes the desired predicates that we want to be true.

In Fig. 2, we show a general representation of a planning task in PDDL. The domain definition (Fig. 2a) is then a general model while the problem definition (Fig. 2b) is a specific problem instance.

```
(define (domain <domain name>)
    <PDDL code for predicates>
    <PDDL code for first action>
    [...]
    <PDDL code for last action>
)
```

(a) Domain definition in PDDL

```
(define (problem <problem name>)
    (:domain <domain name>)
    <PDDL code for objects>
    <PDDL code for initial state>
    <PDDL code for goal specification>
)
```

(b) Problem definition in PDDL

Fig. 2. Planning task in PDDL.

3.3 Planning Systems

The main purpose of writing a planning model in PDDL is to use planning systems (also known as planners) to solve challenging problems. A planner takes as input a model of the domain and a problem instance. As a result, it generates a plan as an answer to the specified input problem.

In the last 20 years, the automated planning community has developed a variety of state-of-the-art planners able to scale up to large problems thanks to the use of effective domain-independent heuristics which allow to guide the search. The algorithms embedded in planning systems search for paths in the search space through different search strategies. They also have properties such as time and memory complexity, completeness and optimality.

From the performance results in past planning competitions [16], we can highlight the planners SymBA*-2 [15], YAHSP3 [17], BFS(f) [10] and IBaCoP2 [4]. Since it is not the aim of this work to make a complete review of neither all planning systems nor all the different approaches, we leave it to the interested reader to consult the extensive literature available [3,6,7].

4   AI Planning for Smart Grids: Initial Investigation

Smart electrical grids play a major role in energy transition but raise important software problems. Some of them can be efficiently solved by AI techniques [2]. In particular, the increasing use of distributed generation based on renewable energies (wind, photovoltaic, among others) leads to the issue of its integration into the distribution network.

Some methods have been used to automatically build target architectures to be reached within a given time horizon (of several decades) capable of accommodating a massive insertion of distributed generation while guaranteeing some technical constraints [1]. However, these target networks may be quite different from the existing ones and therefore a direct mutation of the network would be too costly. It is therefore necessary to define the succession of works year after year to reach the target, i.e. the intermediate grids.

As previously said, the task of determining the intermediate grids that allow to reach the target grid is done by hand. Our assumption is that AI planning could fit this task very well. First, AI planning is beneficial to address problems subject to continuous change (e.g. integration of renewable energies, management of the load caused by electric vehicles, etc.). Indeed, writing and modifying a planning model is less costly and time consuming than implementing a solver and adapting it with each new change. Second, planning tools are indicated when the problem is to find a solution path in a large transition system. In our case, the intermediate grids are defined as the path allowing the transition from the initial grid to the target grid. Finally, AI planning provide a wide open space of promising algorithmic possibilities.

Running Example. In Fig. 3, we present a distribution network of six nodes that will allow to illustrate our approach. It consists of a primary substation (square) connected to another primary substation via main feeders (lines) supplying secondary substations (small and big circles). The nodes have been numbered from 0 to 7 for ease of reading when we refer to them.

We recall that secondary substations with one normally closed switch (NCS) and one Normally Open Switch (big circle) provide a radial grid and allow the grid to be reconfigured in case of a fault. In the initial grid (see Fig. 3a) NOS are located in nodes 1 and 4, while in the target grid (see Fig. 3b) they are located in nodes 3 and 4. Additionally, we consider the following open lines (dashed line):

- Initial Grid:
  - NOS in node 1 is open towards node 2.
  - NOS in node 4 is open towards node 5.
- Target Grid:
  - NOS in node 3 is open towards node 2.
  - NOS in node 3 is open towards node 5.

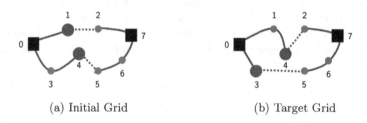

(a) Initial Grid   (b) Target Grid

Fig. 3. Running example of a distribution network.

4.1 Modelling of the Distribution Network Knowledge

Modelling Guidelines. In this initial investigation, we only address the dynamics of line connections. We consider grids in urban environments. We assume that we are provided with an initial grid and a target grid. The target grid should be reachable from the provided initial grid, i.e. both grids have the same primary and secondary substations and differ only by their connections and their normally open switches. Modelling guidelines are focused in finding the transition from the initial grid to the target grid, i.e. the intermediate grids. Each one of the them must be valid in itself. This means that an intermediate grid must follow the rules below:

- A primary substation cannot be isolated, i.e. there must always be at least one secondary substation connected to it.
- A secondary substation cannot be isolated, i.e. there must always be a path that connects it to a primary substation.

- There must be at least (but not more than) one NOS on each set of lines (i.e. main feeder) that connect two primary substations together. For example, in Fig. 3a the main feeder that connects primary substations 0, 7 with secondary substations 1, 2 has only one NOS on node 1.
- A primary substation can be directly connected to one or more secondary substations. For example, the primary substation at node 0 could have another secondary substation connected to it.
- A secondary substation can be directly connected to at most two other, primary or secondary, substations. For example, the secondary substation at node 6 (see Fig. 3a) already has two connections (with nodes 5 and 7) and therefore cannot have more. This constraint has been arbitrarily chosen and can be modified by slightly changing the model.

Finally, following the intermediate grids, one by one, from the initial grid must result in the target grid.

Formalisation into PDDL. We address the task of determining the intermediate grids by translating it to an Automated Planning task. The real challenge is to write the domain model. We therefore need to define the main elements of a PDDL planning task (see Sect. 3.2) in the distribution network domain. These elements are:

- Predicates:
  - Is there an open line between x substation and y substation? (PDDL-Code 1.1 line 5)
  - Is x substation connected to y substation? (line 6)
  - Is x substation feed by p primary substation? (line 7)
  - Is s secondary substation free to receive another connection? (line 8)
  - Is x substation mutable? (line 9)
- Actions/Operators:
  - A line between two secondary substations can be removed if there exists a normally open switch between them. For example, we can close the NOS at 4 and remove the line between nodes 4 and 5. (PDDL-Code 1.1 line 11)
  - A line between a secondary substation and a primary substation can be removed if there exists a normally open switch between them. (line 23)
  - A secondary substation can change the direction of its normally open switch. For example, the NOS of the secondary substation at node 4 is open towards node 5 but it can be closed here and be open for the line between 3 and 4. (line 34)
  - Two secondary substations can be connected and a NOS can be placed in one of them towards the other substation, if each of them can still receive an additional connection. (line 47)
  - A secondary substation can be connected to a primary substation and a NOS can be placed in the secondary substation the primary substation, if the secondary substation can still receive an additional connection. (line 60)

Notice that each action has a set of precondition which determines if they can be applied in a given state. If the action can be applied, it alters the set of true facts according to the action's effects. We provide in PDDL-Code 1.1, the complete definition of predicates and actions.

```
1   (define (domain DISNET)
2     (:requirements :typing)
3     (:types primary secondary - substation)
4     (:predicates
5       (open_line ?x - substation ?y - substation)
6       (connected ?x - substation ?y - substation)
7       (feed ?x - substation ?p - primary)
8       (free_for_connection ?s - secondary)
9       (mutable ?x - substation)
10    )
11    (:action remove_openLine_secTosec
12      :parameters (?s1 - secondary ?s2 - secondary)
13      :precondition (and
14          (open_line ?s1 ?s2) (connected ?s1 ?s2)
15          (mutable ?s1) (mutable ?s2))
16      :effect (and (not(connected ?s1 ?s2))
17          (not(connected ?s2 ?s1))
18          (not(open_line ?s1 ?s2))
19          (not(open_line ?s2 ?s1))
20          (free_for_connection ?s1)
21          (free_for_connection ?s2)
22          ))
23    (:action remove_openLine_secTopri
24      :parameters (?s1 - secondary ?p1 - primary)
25      :precondition (and
26          (open_line ?s1 ?p1) (connected ?s1 ?p1)
27          (mutable ?s1))
28      :effect (and (not(connected ?s1 ?p1))
29          (not(connected ?p1 ?s1))
30          (not(open_line ?s1 ?p1))
31          (not(open_line ?p1 ?s1))
32          (free_for_connection ?s1)
33          ))
34    (:action change_switch_connection
35      :parameters (?s1 - secondary ?s2 - substation
36      ?s3 - substation ?p1 - primary ?p2 - primary)
37      :precondition (and
38          (open_line ?s1 ?s2) (connected ?s1 ?s2)
39          (connected ?s3 ?s1) (feed ?s3 ?p1)
40          (feed ?s1 ?p1) (feed ?s2 ?p2))
41      :effect (and (not(open_line ?s1 ?s2))
42          (not(open_line ?s2 ?s1))
43          (not(feed ?s1 ?p1))
44          (open_line ?s1 ?s3)
45          (open_line ?s3 ?s1)
```

```
46          (feed ?s1 ?p2)))
47     (: action open_line - connect_secTosec
48       : parameters (?s1 - secondary ?s2 - secondary)
49       : precondition (and
50           (free_for_connection ?s2) (mutable ?s2)
51           (free_for_connection ?s1) (mutable ?s1))
52       : effect (and
53           (not(free_for_connection ?s2))
54           (not(free_for_connection ?s1))
55           (open_line ?s1 ?s2)
56           (open_line ?s2 ?s1)
57           (connected ?s2 ?s1)
58           (connected ?s1 ?s2)
59           ))
60     (: action open_line - connect_secTopri
61       : parameters (?s1 - secondary ?p1 - primary)
62       : precondition (and
63           (free_for_connection ?s1) (mutable ?s1))
64       : effect (and
65           (not(free_for_connection ?s1))
66           (open_line ?s1 ?p1)
67           (open_line ?p1 ?s1)
68           (connected ?p1 ?s1)
69           (connected ?s1 ?p1)
70           )))
```

PDDL-Code 1.1. Definition of the distribution network domain.

As previously said, writing the domain model is challenging. Additionally, it is also necessary to verify that the domain is expressive enough to specify a problem. We therefore need to write the problem specification. To illustrate this, we have the following elements for the running example:

- Objects:
  - Two primary substations. $P1$ refers to node 0 and $P2$ refers to node 7.
  - Six secondary substations. S_i refers to node i where $i \in 1, 2, 3, 4, 5, 6$.
- Initial state:
  - A *connected(x,y)* predicate and its mirror, for each two connected substations x and y in the initial grid.
  - A *open_line(x,y)* predicate and its mirror, for each open line between two substations x and y in the initial grid.
  - A *feed(x,p)* predicate, for each substation x feed by a primary substation p in the initial grid.
  - A *mutable(x)* predicate, for each substation x in the initial grid that changes in comparison to the target grid. For example, the secondary substation at node 1 is *mutable* since in the initial grid it is connected to node 2 but in the target grid it is not. On the contrary, the secondary substation at node 6 is *not mutable* since in the initial and in the target grid it is connected to the same nodes (i.e. nodes 5 and 7).

– Goal specification:
  • A *connected(x,y)* predicate and its mirror, for each two connected substations x and y in the target grid.
  • A *open_line(x,y)* predicate, for each open line between two substations x and y in the target grid.
  • A *feed(x,p)* predicate, for each substation x feed by a primary substation p in the target grid.

We provide in PDDL-Code 1.2, the complete definition of the initial state and the goal specification for the running example in Fig. 3.

```
1   (define (problem disnet001) (:domain disnet)
2   (:objects
3     P1 P2 - Primary
4     S1 S2 S3 S4 S5 S6 - Secondary)
5   (:init (mutable S1) (mutable S2) (mutable S3) (mutable S4)
6     (mutable S5) (connected P1 S1) (connected S1 P1)
7     (connected S1 S2) (connected S2 S1) (connected P2 S2)
8     (connected S2 P2) (connected P1 S3) (connected S3 P1)
9     (connected S3 S4) (connected S4 S3) (connected S4 S5)
10    (connected S5 S4) (connected S5 S6) (connected S6 S5)
11    (connected P2 S6) (connected S6 P2) (feed P1 P1)
12    (feed S1 P1) (feed S3 P1) (feed S4 P1) (feed P2 P2)
13    (feed S2 P2) (feed S6 P2) (feed S5 P2) (open_line S1 S2)
14    (open_line S2 S1) (open_line S4 S5) (open_line S5 S4))
15  (:goal (and (connected P1 S1) (connected S1 P1)
16    (connected S1 S4) (connected S4 S1) (connected S4 S2)
17    (connected S2 S4) (connected S3 P1) (connected P1 S3)
18    (connected S3 S5) (connected S5 S3) (connected S6 S5)
19    (connected S5 S6) (connected P2 S2) (connected S2 P2)
20    (connected P2 S6) (connected S6 P2) (open_line S4 S2)
21    (open_line S2 S4) (open_line S3 S5) (open_line S5 S3)
22    (feed P1 P1) (feed S1 P1) (feed S4 P1) (feed S3 P1)
23    (feed P2 P2) (feed S2 P2) (feed S6 P2) (feed S5 P2))))
```

PDDL-Code 1.2. Definition of a distribution network problem.

4.2 Experimental Results

The modelled domain and each problem instance were fed to the DFS+ [11] planner. The planner was run with a timeout of 10 s and 2 GB of RAM.

The planner has always successfully found a plan for each of our problem instances. In Table 1, we present for each tested problem instance the parameters of the initial and the final grid (number of primary and secondary substations, the number of main feeders), the number of steps in the obtained solution and the time, in seconds, taken by the AI planner to obtain it.

Obtained plans describe the set of lines to be built and deconstructed from the initial grid until the target grid is reached. In PDDL-code 1.3, we show the seven step solution found by the planner for the problem instance disnet001 (i.e. our running example). In Fig. 4, we provide a graphical representation of this

Table 1. Description of the problem instances and obtained results.

Problem instance	Primary substations	Secondary substations	NOS initial grid	NOS target grid	Plan length	Search time (s)
disnet001	2	6	2	2	7	0.002
disnet002	2	6	2	1	13	0.003
disnet003*	2	10	2	2	35	0.07
disnet004	2	10	2	3	23	0.021
disnet005	2	20	4	4	41	1.75

plan, i.e. the intermediary grids and actions from the initial grid until reach the target grid.

```
1  (remove_openLine_sectosec s5 s4)
2  (remove_openLine_sectosec s2 s1)
3  (open_line-connect_sectosec s1 s4)
4  (change_switch_connection s4 s1 s3 p1 p1)
5  (remove_openLine_sectosec s4 s3)
6  (open_line-connect_sectosec s3 s5)
7  (open_line-connect_sectosec s2 s4)
```

PDDL-Code 1.3. Plan for the problem instance of the running example (disnet001).

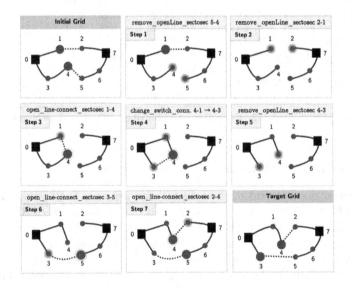

Fig. 4. Graphical representation of the plan in PDDL-Code 1.3.

5   Discussion and Conclusion

Our results clearly show that AI planning is a very promising way to solve smart grid engineering problems. The AI planner, provided with our planning model, was able to solve all problem instances very quickly (less than two seconds). In this regard, we asked a smart grid expert to solve the problem instance disnet003 without the help of AI planning. This problem instance consist of two primary substations, ten secondary substations and two feeders (see Table 1). The solution given by the expert is 21 actions long and was found in about one hour (3600 s). In contrast, the solution given by the AI planner is 35 actions long and was found after 0.07 s.

Given the time taken by the expert to solve an easy instance, we can intuitively think that increasing the number of substations could become problematic. Indeed, if we only consider from the space of all possible grids, those where all secondary substations are part of a single main feeder then we can give a lower bound on the number of possible grid configurations. If s is the number of secondary substations then we have a lower bound in $O(s!)$ which grows over-exponentially. To illustrate this, in Fig. 5, we show a representation of the explored state space for the instance disnet001 with six secondary substations in comparison with the instance disnet003 with ten secondary substations. We go from 568 visited states (Fig. 5a) to 2795 (Fig. 5b), and from a height of the search tree of 11 to 35. Because of this fast growing complexity, this task can quickly become intractable for human experts.

Regarding the plan length, our solution (35 actions) is longer than the one given by the expert (21 actions). This could be a disadvantage for AI planning.

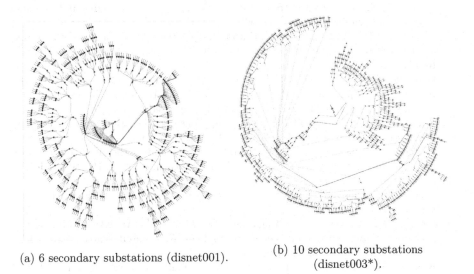

(a) 6 secondary substations (disnet001).

(b) 10 secondary substations (disnet003*).

Fig. 5. Representation of the explored state space for two different problem instances.

However, some of the actions given by the AI planner can be easily simplified to a single action and, therefore, our plan can be shortened. In particular, this is the case for the action `change_switch_connection` since the planner acts locally to ensure that each intermediate grid is valid. For example, let us imagine that we have the grid 0-1=:2-3-4-5 where '-' denotes a connection, '=' denotes an open line and ':' denotes a NOS. If the solution is 0-1-2-3-4:=5, the AI planner has to go through each intermediate change, i.e. 0-1=:2-3-4-5, 0-1-2:=3-4-5, 0-1-2-3:=4-5 and 0-1-2-3-4:=5. This can be solved in a plan post-processing step with a simple script. By applying such a post-processing step, we obtained a shortened plan of 25 actions. In addition, the plan could be further optimised to avoid unnecessary actions.

Finally, AI planning has already demonstrated its applicability to build a system for defining substation voltage targets for the Grendon substation, near London, England [2]. Authors proposed a numeric, non-temporal planning model. Our results and that work only encourage us to continue exploring AI planning for smart grids.

The objective of further work is therefore to mutate an initial network towards a target network while minimising the total cost and respecting technical constraints.

References

1. Alvarez, M.C., et al.: Distribution network long term planning methods comparison with respect to DG penetration. In: I-SUP 2008 (Innovation for Sustainable Production) (2008)
2. Bell, K., Coles, A., Fox, M., Long, D., Smith, A.: The application of planning to power substation voltage control. In: ICAPS Workshop on Scheduling and Planning Applications (SPARK), pp. 1–8 (2008)
3. Castellanos-Paez, S.: Learning routines for sequential decision-making. Ph.D. thesis, Université Grenoble Alpes, Grenoble (2019). https://tel.archives-ouvertes.fr/tel-02513236
4. Cenamor, I., De La Rosa, T., Fernández, F., et al.: IBACOP and IBACOP2 planner. IPC 2014 planner abstracts pp. 35–38 (2014). https://helios.hud.ac.uk/scommv/IPC-14/repository/booklet2014.pdf. Accessed 05 Jan 2021
5. Fikes, R., Nilsson, N.: STRIPS: A new approach to the application of theorem proving to problem solving. Artif. Intell. **3–4**(2), 189–208 (1971)
6. Geffner, H., Bonet, B.: A concise introduction to models and methods for automated planning. Synth. Lect. Artif. Intell. Mach. Learn. **8**(1), 1–141 (2013). Morgan & Claypool Publishers
7. Ghallab, M., Nau, D., Traverso, P.: Automated planning: theory and practice. Elsevier (2004)
8. Haslum, P., Lipovetzky, N., Magazzeni, D., Muise, C.: An introduction to the planning domain definition language. Synth. Lect. Artif. Intell. Mach. Learn. **13**(2), 1–187 (2019). Morgan & Claypool Publishers
9. Khator, S.K., Leung, L.C.: Power distribution planning: a review of models and issues. IEEE Trans. Power Syst. **12**(3), 1151–1159 (1997). IEEE
10. Lipovetzky, N., Geffner, H.: Width and serialization of classical planning problems. In: ECAI 2012, pp. 540–545. IOS Press (2012)

11. Lipovetzky, N., Geffner, H.: Width-based algorithms for classical planning: new results. In: ECAI 2014, pp. 1059–1060. IOS Press (2014)
12. McDermott, D., et al.: PDDL-the planning domain definition language. Technical Report (1998)
13. McDermott, D.M.: The 1998 AI planning systems competition. AI Mag. **21**(2), 35 (2000)
14. Temraz, H.K., Quintana, V.H.: Distribution system expansion planning models: an overview. Electr. Power Syst. Res. **26**(1), 61–70 (1993). Elsevier
15. Torralba, A., Alcázar, V., Borrajo, D., Kissmann, P., Edelkamp, S.: SymBA*: A symbolic bidirectional A* planner. In: International Planning Competition, pp. 105–108 (2014). https://helios.hud.ac.uk/scommv/IPC-14/repository/booklet2014.pdf. Accessed 07 Jan 2021
16. Vallati, M., Chrpa, L., Mccluskey, T.L.: What you always wanted to know about the deterministic part of the international planning competition (IPC) 2014 (but were too afraid to ask). Knowl. Eng. Rev. **33** (2018). Cambridge University Press
17. Vidal, V.: YAHSP3 and YAHSP3-MT in the 8th international planning competition. In: Proceedings of the 8th International Planning Competition (IPC-2014), pp. 64–65 (2014). https://helios.hud.ac.uk/scommv/IPC-14/repository/booklet2014.pdf. Accessed 07 Jan 2021

Artificial Intelligence Application for Crude Distillation Unit: An Overview

Václav Miklas[1] , Michal Touš[1] , Vítězslav Máša[1](✉) , and Sin Yong Teng[2]

[1] Institute of Process Engineering and NETME Centre, Brno University of Technology, Technická 2896/2, 616 69 Brno, Czech Republic
masa@fme.vutbr.cz

[2] Institute for Molecules and Materials, Radboud University, Heyendaalseweg 135, 6525 AJ Nijmegen, The Netherlands

Abstract. Artificial intelligence (AI) with its efficiency for complex systems is growing in popularity in many engineering fields. The ability of an AI method to be successfully applied is highly dependent on the previous research, which makes knowledge sharing within and across fields extremely valuable. This work focuses on crude distillations units (CDU), whose energy optimization has been a tremendous challenge because of its complexity. The presented overview shows that soft sensors are the most common application of artificial intelligence for a CDU, although a number of recent publications focus on optimization problems. The approaches for optimization are very diverse, which makes them hardly applicable in the current engineering practice. This work provides a guideline for selecting the right method, but also addresses the fact that different methods excel at different problems and with different data set sizes. For neural networks (NN), this further depends on their architecture and hyperparameter adjustment. This urges future research, whose goal could be a workflow that would automatically adapt methods and perform parameter tuning with minimum user input.

Keywords: Crude Distillation Unit (CDU) · Artificial Intelligence (AI) · Machine Learning (ML) · Neural Network (NN) · Sustainability

1 Introduction

Artificial intelligence is a dynamically developed field, whose significance grew within various industries over the last few decades. While the origins of artificial intelligence (AI) and early neural networks (NN) can be traced 1940s [1], the breakthrough did not happen until the new millennium when rapid hardware development allowed for practical implementation [2].

Nowadays, AI has some extent of successful implementation in virtually all aspects of everyday life, including engineering. Some highly-cited published engineering applications range from optimization of constrained design problems [3], through multi-agent and holonic manufacturing systems [4], risk analysis and maintenance [5], to water quality prediction [6].

E. Mercier-Laurent and G. Kayakutlu (Eds.): AI4KMES 2021, IFIP AICT 637, pp. 156–168, 2022.
https://doi.org/10.1007/978-3-030-96592-1_12

Some of the important engineering challenges to address using artificial intelligence are the reduction of energy consumption and related emissions, as well as sustainability of processes in general. Whereas a Scopus search "TITLE-ABS-KEY (artificial AND intelligence) OR TITLE-ABS-KEY (machine AND learning) OR TITLE-ABS-KEY (neural AND network)" returns almost 1.2 million results, mere 3,400 documents mention "sustainability" in their title, abstract or keywords.

One of the industries where the abovementioned sustainability-related keywords are relevant is petroleum refining, whose global primary energy consumption is estimated to be over 5% [7]. Within oil refineries, the key equipment which is also related to the highest energy consumption is crude distillation unit (CDU), which might use around one-third of the refinery primary energy demands [8].

Chemical engineering, including oil refining, somewhat lagged behind the general increasing trend in artificial intelligence implementation at the beginning of the millennium. This is demonstrated in Fig. 1. However, the last couple of years showed an unprecedented growth, from around 200 publications indexed in Scopus per year between 2015 and 2017 to 700 records in 2020.

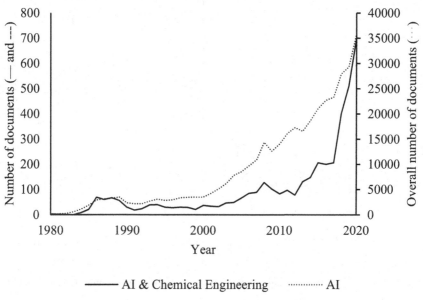

—— AI & Chemical Engineering   ············ AI

Fig. 1. All results for the "artificial intelligence" Scopus search compared to the same search narrowed down to the subject areas "Chemical Engineering".

This work presents an overview of artificial intelligence applications for the CDU over the last decade, which, especially in its second half, recorded such a rapid increase in publishing activity within chemical engineering. However, only a limited number of publications deal with the energy intensity of the process, as will be shown below. The following chapters present the CDU process (2), literature search methodology (3) and the main findings in the literature (4).

2 Process Description

A crude distillation unit (CDU) represents an essential separation step and the first of the fundamental unit operations in any petroleum refinery. Its main purpose is to separate the crude oil feed, a mixture of hydrocarbons with a wide range of boiling points, into semi-finished or final products based on their respective volatility. A common process layout of such a unit is shown in Fig. 2.

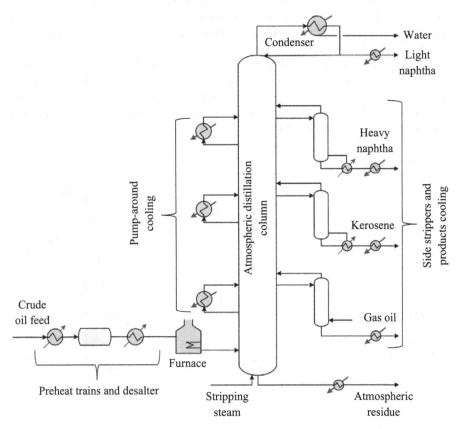

Fig. 2. A simplified process diagram of a typical crude distillation unit; adapted from [17].

Before entering the *atmospheric distillation column* (Fig. 2), the crude oil feed is preheated in a heat exchanger network or series of networks called *preheat trains* (Fig. 2). Heat is recovered from relatively hot streams elsewhere in the process. This heat integration allows for a higher energy efficiency of the unit.

The remainder of the required heat input is provided in a *furnace* (Fig. 2), at the outlet of which the crude oil is partially vaporized and enters the flash zone of the *column*. A relatively steep temperature gradient exists within the *column*, with the lowest temperature at the top. The gradient is caused and controlled by the overhead *condenser* (Fig. 2) as well as *pump-around cooling* (Fig. 2) along the column. The temperature at a specific point in the column is dependent on the vapor-liquid contact between the internal reflux stream and vapor flowing up from the flash zone. The temperature profile determines the position of side draws where different products can be extracted from the main *column.*

Side strippers (Fig. 2) are used to remove light ends (light hydrocarbons) from the products. This is necessary to adjust products' distillation curves and other properties according to the specification. Stripping is done by *stripping steam* (Fig. 2) injection, which is also used at the bottom of the main column, or via a reboiler.

The heaviest fraction, *atmospheric residue* (Fig. 2), leaves at the bottom of the column and is further processed in vacuum distillation and/or other processes. The number and character of lighter liquid products from the side strippers and condenser may vary, but these are commonly (from heaviest to lightest) *gas oil, kerosene, heavy naphtha,* and *light naphtha* (Fig. 2). The light ends and other non-condensable gases are separated in the *condenser* (this stream is omitted in Fig. 2 for simplicity).

Even the significantly simplified illustration and description of the unit demonstrates the high complexity of the system. This work's authors have recently dealt with a CDU system in a similar scope but in a high level of detail. The number of manipulated variables in this case study, already narrowed according to energy optimization purposes, was close to 50. The number of all variables, including dependent and nonrelevant variables, easily goes into hundreds for a typical CDU.

From the authors' experience, even conventional process simulation as the most popular method to design, optimize and troubleshoot crude distillation units can fail to obtain results, given an inadequate combination of manipulated variables. While this is not the method's drawback and often simply indicates an infeasible operating point, it calls for application of unconventional and progressive methods, effective for high-dimensional data and complex problems. For this reason, the following chapters explore the existing applications of artificial intelligence to CDU.

3   Literature Search Methodology

The goal of the methodology is to find relevant records, in which artificial intelligence (AI) is applied to a crude distillation unit (CDU) regardless of the specific purpose. To perform such search, four keyword phrases are used:

- Crude distillation (the investigated process), AND
- Artificial intelligence (as the most general term), OR
- Machine learning (the largest subcategory of AI), OR
- Neural network (the most pronounced method of AI),

The abstract and citation database Scopus is used for the search. The phrases above are looked for within titles, abstracts and keywords. Also, the year range is limited to 2011–2021. The initial search is narrowed by excluding the results which could not be retrieved (reason A), are not available in English (reason B), are not related to crude distillation or artificial intelligence (reason C), are not individual contributions (e.g., conference reviews; reason D), or use the same method or have the same purpose as in the authors' previous work (reason E).

Within the final literature set included in the overview, four main data items are collected wherever available:

– Used AI and non-AI methods,
– Purpose of the work,
– Size and/or division of a data set into training, test and validation sets if applicable,
– Affinity to the topics of sustainability, energy and emissions.

The described multi-step procedure will lead to a relevant set of documents that can be analyzed with respect to qualitative and quantitative parameters that allow for discussion, recognition of well-covered aspects of the process and identification of potential future directions that are not as well-developed.

The main goal, however, is to provide an insight to engineers and researchers dealing with the same tasks that can already be found in the literature. Instead of "reinventing the wheel", they can rely on the methods that have already been created and established.

4 Main Findings in the Literature

The initial search resulted in a group of 45 records. Three following checks, which can be described as a) availability (exclusion reasons A and B), b) eligibility (exclusion reasons C and D), c) uniqueness (exclusion reason E), resulted in a final set of 16 documents that were thoroughly examined and included in the final overview. See Fig. 3 for details about how many documents were excluded in each step.

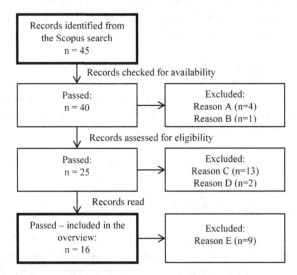

Fig. 3. Step-by-step selection of the final literature set from the initial search.

4.1 Categorization

Based on their purpose, the 16 items can be divided into the following categories:

1. Control of the CDU (Table 1)

   a. Soft sensors – 6 records;
   b. Predictive control – 1 record;

2. Optimization of the CDU (Table 2)

   a. Heat, energy and/or exergy budget – 4 records;
   b. Cost – 2 records (1 design, 1 operation);
   c. Non-specific opt. (1 record);

3. Other (Table 3)

   a. Property estimation – 1 record;
   b. Risk management – 1 record.

The results of the investigation in Tables 1, 2 and 3 show what methods the authors used, the purpose of their efforts, a condensed description of their work, the size and distribution of their data set, and links to the topics of energy, emissions or sustainability (EES).

The application for soft sensors is prevalent, mostly for quantities that cannot be measured continuously, or where the related analysis is time- or cost-intensive (cut point temperature, dry point, vapour pressure).

The optimization purposes are almost as common as soft sensors and very often focus on or at least account for energy demands. However, the variety of used approaches make it uneasy to select the right methods for engineering applications. Authors often claim their approach is superior to others, and rightfully so – for the given problem, data set size, selected network architecture and hyperparameters.

In a sense, both implementations from the Other category (Table 3) can be understood as soft sensors, because both TBP data [10] and corrosion rate [12] are quantities that are costly to measure and not available in real time. They are excluded from the first category because of the limited relation to the process unit and the lack of a tangible location where the soft sensor would predict the quantities.

4.2 Discussion

Feedforward neural networks (FNN), and multilayer perceptron (MLP) which belongs to this category, were most dominant across all the applications. There was, however, a slight bias introduced to the search since NN were the only AI method that was explicitly searched for.

One common feature is the lack of data, which is often compensated using process simulation (Aspen HYSYS [14, 16–18, 20], Aspen Plus [19, 21], PRO/II [13]) to generate a data set mimicking real operating data. While the motivation is sometimes the

Table 1. Overview of the records related to the process control.

Methods	Purpose	Description	Training / Test / Validation	EES	Ref.
1a. Soft sensors					
-K-means -K-nearest neighbors (K-NN) -Random forest (RFc) -Support Vector Classification (SVC) -multilayer perceptron (MLP)	soft sensor for the "needle penetration" analysis (UNE-EN 1426:2015)	-GridSearch used to optimize hyper parameters for each method (e.g. number of neurons in 1 hidden layer of the MLP) -94 inputs -1 output -K-means best for this application	90 % / 10 % (268 samples)	No	[9]
-multiple linear regression (MLR) -linear neural network (LNN) -multilayer perceptron (MLP) -radial basis function (RBF)	prediction of Reid vapor pressure (RVP) of the light naphtha product	-96 data samples -number of neurons and learning algorithms tested to find the best-performing combination -MLP (6-4-1) and RBF (6-7-1) best results	50 % / 25 % / 25 %	No	[10]
-deep belief network (DBN) -restricted Boltzmann machines (RBM)	soft sensor (ASTM 95% cut point temperature)	-DBN consists of a series of RBMs -16 inputs -1 output -architecture 16/20/16/1 -compared and superior to SVM, PLS, NNPLS	251/100	No	[11]
-multilayer perceptron (MLP) -least absolute shrinkage and selection operator (LASSO)	soft sensor for the estimation of the kerosene D95	-compared to sequential backward multiplayer perceptron (SBS-MLP), normalized mutual information feature selection (NMIFS) and ELM (Elman network) -25 variables -25-4-1 architecture	240 / 121	No	[12]

(continued)

Table 1. (*continued*)

Methods	Purpose	Description	Training / Test / Validation	EES	Ref.
a number of statistical learning methods (partial least squares, neural networks, relevance vector machine, Bayesian linear regression, ridge regression)	soft sensor (95% cut point temperature)	-the process simulator PRO/II to generate data -relevance vector machine (RVM) was the best for this application (high accuracy, >10 times faster than NN, high generalization capacity)	225 (50) / 75	Yes	[13]
-bootstrap aggregated neural networks (BANN) -bootstrap aggregated partial least squares (BAPLS)	estimation of kerosene dry point with varying crudes	-16 inputs (selected out of 50) -Aspen HYSYS to generate simulated data, random noise added to the data -intentionally limited data set to represent limited dry point analyses -BANN for oil classification, BAPLS (partial least square) for the dry point prediction	200 / 70 / 87	No	[14]
1b. Predictive control					
- feedforward NN (FNN) -stochastic optimization algorithm Adam (training)	model predictive control for a CDU	-CVXOPT (quadratic programming solver used for data generation and as a benchmark) -architecture: 1072 / (1664–1920) × 3 hidden layers / 32 -NN milliseconds, QP solver tens of seconds, negligible performance loss	1500 epochs	No	[15]

Table 2. Overview of the records related to process optimization.

Methods	Purpose	Description	Training / Test / Validation	EES	Ref.
2a. Heat, energy and/or exergy budget optimization					
-neural network (NN) -Taguchi method -genetic algorithm (GA)	optimize cut point temperatures and specific energy, deal with feed uncertainty	- ANN predicts products' cut points based on the feed composition (reduces calculation time) - 3 hidden layers (13, 17, 16 neurons) - Aspen HYSYS used for energy demands calculation - Taguchi and GA optimize cut point temperatures	85 % / 15 % (192 data points)	Yes	[16]
-feedforward NN (FNN) -simulated annealing (SA)	retrofit of heat exchanger network (HEN) for optimizing CDU operation	-1 hidden layer -scenarios generated in Aspen HYSYS -profit is the objective function to maximize	70 % / 15 % / 15 % (800 data points)	Yes	[17]
-bootstrap aggregated neural networks (BANN)	maximization of exergy efficiency of a CDU	-BANN contains 30 NNs -Aspen HYSYS used for operating data generation	50 % / 30 % / 20 %	Yes	[18]
-evolutionary algorithm assisted by adaptive surrogate functions (EASF) -radial basis function (RBF) -approximation error fuzzy control strategy (AEFCS)	maximize distillation heat supply from reflux streams	-Aspen Plus – rigorous model -DEA (differential EA) – optimization -RBF – surrogate model, three layers -AEFCS – model management -ba	N/A	Yes	[19]

(continued)

Table 2. *(continued)*

Methods	Purpose	Description	Training / Test / Validation	EES	Ref.
2b. Cost optimization					
-neural network (NN) -support vector machine (SVM) -genetic algorithm (GA)	design of a crude distillation unit with optimized cost	-Aspen HYSYS used to build a surrogate model (ANN) while feasibility constraints are generated using a SVM to optimize the column configuration -annual cost as an objective function -18 inputs -46 output (divided into multiple NNs) -one hidden layer (10 neurons)	-70 % / 15 % / 15 % (NN) -75 % / 25 % (SVM)	Yes	[20]
-wavelet neural network (WNN) -line-up competition algorithm (LCA)	economic optimization of a CDU under prescribed constraints	-WNN predicts the CDU behavior, LCA applied to the constrained optimization problem -one hidden layer, 30 neurons -Aspen Plus used for the rigorous model -WNN superior to RBFNN and BPNN, LCA superior to GA and PSO	350/150	Yes	[21]
2c. Non-specific optimization					
efficient dropout neural network (EDN)	optimization problem of crude oil distillation units	-advantageous for high-dimensional multi-objective expensive optimization problem -up to 100 decision variables and 20 objectives -two hidden layers -sensitivity analysis on parameters	various	No	[22]

Table 3. Overview of the other records from the literature search.

Methods	Purpose	Description	Training/Test/ Validation	EES	Ref.
3a. Property estimation					
Sure independence screening and sparsifying operator (SISSO)	ML used for the creation of a simple equation (TBP data to products' relative amounts)	Laboratory distillation of crude oil	–	No	[23]
3b. Risk management					
-Feedforward NN (FNN) -Multiple linear regression analysis (MLRA)	Corrosion prediction and the related risk management	-11 inputs -Two hidden layers -3 outputs (pH, chloride and salt content in crude) -MLRA used to predict the corrosion rate	60–90%/40–10% (24 points)	No	[24]

simplicity of data generation, this might indicate an underlying problem of the industry's conservatism toward artificial intelligence and limited willingness to share data with researchers.

While there were some attempts to optimize the architecture and parameters in the assessed literature [9, 10], the trial-and-error approach is still most common for this challenge as well as method selection.

5 Conclusions

Artificial intelligence (AI) has been successfully applied to a number of engineering problems and has hugely grown in popularity in chemical engineering lately. In this work, some recent publications which applied artificial intelligence to the most energy-intensive oil refining process, crude distillation, were examined with respect to their purpose, methods, data set character and sustainability ideas.

Based on a multi-step search and selection, 16 records were included in the overview. It was found out that soft or virtual sensors are the most typical implementation of artificial intelligence for this process, while neural networks are the dominant AI method.

The variety of approaches, especially for the optimization of crude distillation units, make AI cumbersome to use for practical engineering problems. An interesting future research direction could focus on the automatic selection of an appropriate algorithm, together with the tuning of hyperparameters associated with machine learning.

Acknowledgement. This research has been supported by the project No. CZ.02.1.01/ 0.0/0.0/16_026/0008413 "Strategic partnership for environmental technologies and energy production", which has been co-funded by the Czech Ministry of Education, Youth and Sports within the EU Operational Programme Research, Development and Education.

References

1. McCulloch, W.S., Pitts, W.: A logical calculus of the ideas immanent in nervous activity. Bullet. Math. Biophys. **5**(4), 115–133 (1943)
2. Schmidhuber, J.: Deep learning in neural networks: an overview. Neural Netw. **61**, 85–117 (2015)
3. He, Q., Wang, L.: An effective co-evolutionary particle swarm optimization for constrained engineering design problems. Eng. Appl. Artif. Intell. **20**, 89–99 (2007)
4. Leitão, P.: Agent-based distributed manufacturing control: a state-of-the-art survey. Eng. Appl. Artif. Intell. **22**, 979–991 (2009)
5. Weber, P., Medina-Oliva, G., Simon, C., Iung, B.: Overview on Bayesian networks applications for dependability, risk analysis and maintenance areas. Eng. Appl. Artif. Intell. **25**, 671–682 (2012)
6. Ömer Faruk, D.: A hybrid neural network and ARIMA model for water quality time series prediction. Eng. Appl. Artif. Intell. **23**, 586–594 (2010)
7. Worrell, E., Bernstein, L., Roy, J., Price, L., Harnisch, J.: Industrial energy efficiency and climate change mitigation. Energ. Effi. **2**(2), 109–123 (2008)
8. Abdulrahman, I., Máša, V., Teng, S.Y.: Process intensification in the oil and gas industry: a technological framework. Chemical Eng. Process. Process Intensification **159**, 108208 (2021)
9. Niño-Adan, I., Landa-Torres, I., Manjarres, D., Portillo, E.: Soft-sensor design for vacuum distillation bottom product penetration classification. Applied Soft Computing **102**, 107072 (2021)
10. Rogina, A., Šiško, I., Mohler, I., Ujević, Ž., Bolf, N.: Soft sensor for continuous product quality estimation (in crude distillation unit). Chem. Eng. Res. Des. **89**, 2070–2077 (2011). https://doi.org/10.1016/j.cherd.2011.01.003
11. Shang, C., Yang, F., Huang, D., Lyu, W.: Data-driven soft sensor development based on deep learning technique. J. Process Control **24**, 223–233 (2014)
12. Sun, K., Huang, S.-H., Wong, D.S.-H., Jang, S.-S.: Design and application of a variable selection method for multilayer perceptron neural network with LASSO. IEEE Trans. Neural Netw. Learn. Syst. **28**, 1386–1396 (2017)
13. Urhan, A., Ince, N.G., Bondy, R., Alakent, B.: Soft-Sensor Design for a Crude Distillation Unit Using Statistical Learning Methods (2018)
14. Zhou, C., Liu, Q., Huang, D., Zhang, J.: Inferential estimation of kerosene dry point in refineries with varying crudes. J. Process Control **22**, 1122–1126 (2012)
15. Kumar, P., Rawlings, J.B., Wright, S.J.: Industrial, large-scale model predictive control with structured neural networks. Comput. Chem. Eng. **150**, 107291 (2021)
16. Durrani, M.A., Ahmad, I., Kano, M., Hasebe, S.: An artificial intelligence method for energy efficient operation of crude distillation units under uncertain feed composition. Energies **11**(11), 2993 (2018)
17. Ochoa-Estopier, L.M., Jobson, M., Smith, R.: Operational optimization of crude oil distillation systems using artificial neural networks. Comput. Chem. Eng. **59**, 178–185 (2013). https://doi.org/10.1016/j.compchemeng.2013.05.030

18. Osuolale, F.N., Zhang, J.: Multi-objective optimisation of atmospheric crude distillation system operations based on bootstrap aggregated neural network models. Comput. Aided Chem. Eng. **37**, 671–676 (2015)
19. Shi, X., Tong, C., Wang, L.: Evolutionary optimization with adaptive surrogates and its application in crude oil distillation. In: 2016 IEEE Symposium Series on Computational Intelligence, SSCI 2016 (2017)
20. Ibrahim, D., Jobson, M., Li, J., Guillén-Gosálbez, G.: Optimization-based design of crude oil distillation units using surrogate column models and a support vector machine. Chem. Eng. Res. Des. **134**, 212–225 (2018)
21. Shi, B., Yang, X., Yan, L.: Optimization of a crude distillation unit using a combination of wavelet neural network and line-up competition algorithm. Chin. J. Chem. Eng. **25**, 1013–1021 (2017)
22. Guo, D., Wang, X., Gao, K., Jin, Y., Ding, J., Chai, T.: Evolutionary optimization of high-dimensional multiobjective and many-objective expensive problems assisted by a dropout neural network. IEEE Trans. Syst. Man, Cyber. Syst. (2021)
23. Giordano, G.F., et al.: Distilling small volumes of crude oil. Fuel **285**, 119072 (2021)
24. Hassanudin, S.N., Aziz, I.A., Jaafar, J., Qaiyum, S., Zubir, W.M.A.M.: Predictive analytic dashboard for desalter and crude distillation unit. In: 2017 IEEE Conference on Big Data and Analytics, ICBDA 2017, pp. 55–60 (2018)

Deep Reinforced Learning for the Governance of a Sample Microgrid

Berkay Gür$^{(\boxtimes)}$ ⓘ and Gülgün Kayakutlu$^{(\boxtimes)}$ ⓘ

Istanbul Technical University, Istanbul, Turkey
{gurb19,kayakutlu}@itu.edu.tr

Abstract. A proximal policy optimization reinforcement learning system is proposed to handle the energy dispatch management of a sample microgrid. The microgrid in question has 3 participants of different classifications, signifying their relative importance and how sensitive they are to energy shortages. The energy within the microgrid is generated by these participants, which are individually equipped with a solar panel and a wind turbine for energy generation, and an energy storage system to store this energy. The environmental conditions, i.e. temperature, wind velocity and irradiation figures of Istanbul are considered to obtain accurate energy generation figures. The microgrid is designed to be grid connected in order to compensate for the uncertainties caused by the weather changes, hence service of the utility is accessed when energy produced & stored cannot respond to the demand. Information security of the participants is respected and to that end, direct energy generation, consumption and storage figures are not supplied to the agent, instead only supply and demand figures are transferred. The agent, using this information, after a period of training, optimizes the system for a reward scheme that rewards energy exports and punishes energy deficits and imports. The results verify the feasibility of proximal policy optimization in managing microgrid energy dispatch.

Keywords: PPO · Microgrid · Energy dispatch

1 Introduction

Climate changes caused by Global Warming reduce the quality of life. Residential response to unexpected meteorological pressures has been the quick proliferation of renewable energy sources such as photovoltaic (PV) panels and wind turbines [1]. The energy generation using the renewable sources are, by their very nature, relatively intermittent and, in the case of PV, limited to the daytime. Compared to conventional energy sources, energy generation with renewable energy sources can be achieved at a smaller scale. This enables energy generation in residential areas such as the rooftops. Application of solar on rooftops also has the effect of reducing transmission and distribution losses associated with energy generation, not to mention that they don't take up any further space [2]. Similarly, wind energy can also be utilized in a similar manner [3].

A problem arising with the growth of renewable energy sources is the contradiction with conventional energy generation methods [4]. The traditional utility grid is ill suited

© IFIP International Federation for Information Processing 2022
Published by Springer Nature Switzerland AG 2022
E. Mercier-Laurent and G. Kayakutlu (Eds.): AI4KMES 2021, IFIP AICT 637, pp. 169–183, 2022.
https://doi.org/10.1007/978-3-030-96592-1_13

to work with the increasing volume of uncoordinated renewable energy generation. One example of this is the phenomenon called the "Duck curve", which refers to the mismatch between the peak demand and renewable energy production. The peak load usually happens at sunset, while renewable energy production reaches its' zenith at mid-evening hours, this causes an energy deficit that can't be satisfied by renewables, requiring the usage of conventional energy sources, and incurring ramp up costs [5].

Hence, microgrids, as a concept, have been designed to cope with the penetration of renewable energy sources [6]. They can be considered as a set of generators, loads and batteries in a single system. This single system can be connected to the larger utility grid at a single point, called a point of common coupling (PCC) and the energy transactions between the utility grid and the microgrid would be managed at that point [7]. It is also possible to manage the microgrid in such a way that enables the microgrid to be independent of the utility grid. This sort of microgrids do not have a PCC since there is no point to be connected to the utility grid. These microgrids are called islanded microgrids. Microgrids with PCC can choose to become islanded by severing the connection between the utility grid and microgrid. This is useful in the cases of catastrophic failure in the utility grid such as massive shortages that may occur due to adverse environmental conditions, such as in the case of Texas power outage of February 2021 [8].

The properties of the grid connected microgrid could be utilized to isolate critical loads for power outages. This paper seeks to use the prioritization property of the microgrid to create a system where the participants of the microgrid are categorized according to their criticality. This would require the microgrid to be constantly supervised for energy deficits and prompt action must be taken to prevent any shortage in the critical loads. To manage this, artificial intelligence could be utilized to automate the monitoring and decision-making aspect of the grid controlling.

To achieve this objective, a solid grasp of microgrids, energy management within the microgrid, utilized methods that exist in the literature should be explored. This requires an organized and methodological survey of the existing literature. Furthermore, this survey would be focused further on the chosen machine learning method, i.e. RL.

2 Literature

RL methods are already applied in the field of energy management of microgrids. The energy management within a microgrid, as explained by Murty and Kumar, encompasses both the supply and demand side management of the microgrid [9]. This can, as the authors explain, can involve managing the demand by incentivizing the usage of energy at particular times, or by managing the dispatch, i.e. the flow of the energy that the microgrid controller can utilize. Qazi et al. [10] proposed a Q-Network system where the supplied energy and the energy present in the energy storage systems are exchanged within an isolated microgrid. Kozlov et al. [11] simulates an isolated microgrid which contains constant loads, solar panels, wind turbines and biomass engines as a Markov Decision Process automates the dispatch of each energy source. In the work of Muriithi and Chowdhury the energy in the energy storage system is traded with the prices and the battery degradation taken into consideration [12]. The microgrid system is modeled as a Markov Decision Process and solved using Q-Learning to minimize costs in a successful manner in different tariffs and seasonal conditions.

One of the tasks that can be achieved with reinforcement learning is the management of energy storage systems within the microgrid system. The article authored by Shang et al. [13] successfully employed Q-Learning with constraints to optimize the load dispatch policies of the energy storage system within a microgrid while taking into account the battery degradation. Research article of Al-Gabalawy [14] applies reinforcement learning along with linear programming to optimize energy transaction decisions of the energy storage systems with both risk averse as well as risk seeking agents. The implementation of energy storage systems are not limited to the use of batteries. Electric vehicles are also explored as in the work of Sadeghi et al. [15] where a novel Bayesian Coalition Game is explored in microgrids with uncertainties due to EV and attempts to reduce the energy losses within the microgrid. Fan et al. [16] consider batteries as well as diesel generator within an isolated microgrid with Deep Deterministic Policy Gradient method to minimize the operating cost and compares the result with transfer learning to prove the superiority of the Deep Deterministic Policy Gradient when used in tandem with Transfer Learning.

Samadi et al. approached this issue from a multi agent perspective [17] where each participant of the microgrid is a seperate agent interacting in a competitive environment. To do this the authors established a Markov Decision Process to find the optimal policy. In this article the authors defined each consumer and energy generator as separate agents and rewards and punishments within the system are the profits and costs of the energy transactions. Fang et al. similarly employs a multi agent approach for each of the microgrid participants where they interact in a double action scheme for the energy transactions and distributes rewards for consumption and generation from various energy sources. Fang et al. takes a similar approach with a multi agent system with renewable energy generation participating in energy transactions through an auction scheme [18]. Their work also takes into account the differing load patterns of industry, commerce and residential loads. Guo et al. [19] modeled the microgrid in a two-level system in a Stackelberg game where the distribution system operator is the leader and the microgrids are the followers and all of the actors in the multi microgrid environment are modeled as a different agent.

Certain design considerations play a huge role in the success of the RL machine learning model. Wu and Wang, in their study, note that the success of a deep reinforcement learning model in the context of microgrids is dependent on the design of the reward structure and badly designed reward schemes could lead to cascading failures [20]. And Yin and Zhang, explain that microgrid operations typically operate on two different time scales, 15-min time steps for transactions and 4 s steps for generation dispatch, which the authors explain can lead to uncoordinated problems [21]. They attempt to create accurate single time steps with particle swarm optimization and reinforcement learning and compare the results with time series generative adversarial networks. The latter is proposed to generate accurate data that functions on a single time step to improve the operation of the microgrids.

The contributions of this paper to the research on the matter of energy dispatch within a microgrid are twofold.

- First is the lack of information to the grid controlling agent to preserve the personal information of the grid participants.

- Second is the prioritization of the grid members according to their importance to their function or social status, thereby securing the needs according to the specifications of the grid controller.

These contributions do not take into usage of PPO in the microgrid environment, which is, to the best knowledge of the authors, unattempted for the problem of energy dispatching with imports and exports.

The rest of the paper is organized as follows. The second section will be a brief introduction to the topic of reinforcement learning and the used sub-category of reinforcement learning, the proximal policy optimization, along with the reasoning behind the usage of the said methods. The third section will be an examination of the microgrid environment. This section will go over the components of the microgrid, how they are represented to the machine learning algorithm, the observation and action spaces of the reinforcement learning algorithm as well as the reward scheme used in the learning process. The next section consists of a discussion of the results. And the last chapter will be the conclusion and how this model could be further improved in the future.

3   Reinforcement Learning

Microgrids, in our case, while using internal energy resources to sustain themselves, remain in connection with the utility grid. This connection allows energy import from the grid at the time of deficit to respond to the demand and export to the grid at the time of excess production. Reinforcement learning is utilized to manage the flow of energy within the microgrids, as well as regulating imported and exported energy.

Reinforcement learning is one the main paradigms of machine learning along with supervised and unsupervised learning [22]. Reinforcement learning is used to carry out trial and error tasks within an environment. How correct are the actions undertaken by the reinforcement learning model is judged by a reward, which itself is a function of the state.

In this paper, the proximal policy optimization algorithm is chosen because the learning process itself is a stochastic process. Proximal Policy Optimization is a series of algorithms that build on the previous policy gradient algorithms that are used to learn the policy directly. Rather than dictating the exploration or exploitation through a set of percentages as in the -greedy method, the policies itself are considered as a probability distribution and each choice made updates the probability, eventually leading to a more stable algorithm than the Q-Learning counterpart [23]. The trade-off is that reaching a suitable solution with policy gradient highly dependent on the step size, if the step size is too low the learning would be very slow and if the step size is too large there may be catastrophic performance implications due to noise [24]. To add on to it, policy gradient algorithms are also sample inefficient, meaning that longer learning time is needed to reach a result compared to the Q-Learning algorithm [25].

Proximal Policy Optimization attempts to solve these issues in policy optimization by utilizing a novel objective function that enables multiple epochs of minibatch updates and reaps the benefits of a trust region policy approach method in a simpler manner with better overall performance [26].

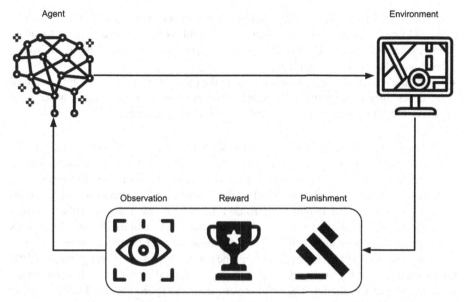

Fig. 1. A graphical representation of the reinforcement learning. Icons made by smalllikeart, Radhe Icon, Prosymbols Premium, Eucalyp and Freepik from www.flaticon.com

4 Microgrid Model

The microgrid, designed for this study, consists of three participants. Each of the participants has its own energy consumption profile, generation, and storage capacities. The participants are also subject to a classification for the criticality (Table 1).

Table 1. Microgrid participant categorization

Participant category	Categorization code
Critical	1
Vulnerable	2
Normal	3

- The critical participants of the microgrid could be described as the participants that cannot experience any energy shortage under any circumstances. Examples to such microgrid loads are hospitals, server rooms or laboratories that may exist in a university campus. This loss of energy on these loads could put the people within the microgrid in danger, put the scientific progress in an academic setting in jeopardy or potentially disrupt business functions in a commercial setting.
- The vulnerable category is the representation of the participants that potentially can't supply or afford to supply their own energy for heating or for electricity. This definition

is similar to the definition established by the EU Energy Poverty Observatory, which states "Energy poverty occurs when a household suffers from a lack of adequate energy services at home" [27]. The lack of energy within these participants would have potential social consequences.

• Lastly, the normal category would represent the participants who can supply or afford to supply their own energy. Although the energy shortage is still undesirable, it can be tolerated for a while and not as critical as the other two users.

All of the participants generate a part of their energy through wind and solar energy. This is calculated on an hourly basis. For solar energy, the Photovoltaic Geographical Information System (PVGIS) is used, which is developed by the European Commission Joint Research Centre and uses the satellite data over Europe and processes the readings into a solar panel power output [28]. To do this, they use data they've collected from solar panels, observed wind velocity, ambient temperature and irradiation. The results of which are logged on an hourly basis for a coordinate over Istanbul, Turkey.

A similar effort is also undertaken from the calculation of wind energy output of the microgrid participants. Energy generated through the wind turbines is dependent on the wind strength of the district. The link between the wind speed and wind turbine output is illustrated as a piecewise function as follows [29].

$$P_{WT,t} \begin{cases} 0 & v_t < v_{ci} \\ P_r \frac{v_t^3 - v_{ci}^3}{v_r^3 - v_{ci}^3} & v_{ci} \leq v_t < v_r \\ P_r & v_r \leq v_t \\ 0 & v_t > v_{co} \end{cases} \tag{1}$$

where v_{ci} and v_{co} are cut in and cut out velocities of the wind turbine respectively. P_r and v_r are rated power and rated velocity of the wind turbine, and v_t is the measured velocity at time of t as processed from the PVGIS dataset for the same coordinate as the solar power output.

The ESS utilized in the system has restrictions in place with regards to the maximum and minimum charge of the energy storage system are implemented as follows.

$$SoC_t \geq SoC_{min} \tag{2}$$

$$SoC_t \leq SoC_{max} \tag{3}$$

The current state of charge SoC_t inside the energy storage system is equal to the summation of the prior state of charge SoC_{t-1} and the incoming or outgoing power $P_{ESS,t}$ multiplied with the efficiency coefficient η.

$$SoC_t = SoC_{t-1} \pm \eta P_{ESS,t} \tag{4}$$

As is in the case of Kim et al. [30] 95% efficiency for the incoming and outgoing energy is implemented, signified by η.

Taking all these calculations into account consumer activities are modeled as a workflow. The first step of this workflow is to assess the hourly consumption, solar and wind generation as described previously. The consumers would judge their power balance, if their generated power is enough to cover for the hourly power consumption of the participants. The hourly balance could be formalized as follows.

$$P_{B,t} = P_{C,t} + P_{PV,t} + P_{WT,t} \tag{5}$$

Where $P_{B,t}$ is the power balance in a given time t, $P_{C,t}$ is the power consumption in a given time t, $P_{PV,t}$ and $P_{WT,t}$ is the solar and wind power generation in a given time t respectively.

If the $P_{B,t}$ is negative, meaning there is an energy deficit between the generated power and consumed power at the given time t, then the deficit will be compensated from the energy storage system. Next, the participant in the deficit will decide whether the energy in the energy storage system is sufficient. If the energy storage systems projected state of charge goes below the minimum level, the energy storage will reach minimum, and the remaining deficit will be satisfied by the microgrid.

Conversely, if the $P_{B,t}$ is positive, that is to say there is a surplus of power in the participant, the participant will initially use this excess to charge its energy storage system. If the energy storage system is full, then the excess will be forwarded to the microgrid for usage or export.

Therein lies one of the novelties of the paper, the reinforcement learning environment does not convey the information of the energy consumption and generation to the agent. The grid manager effectively has no knowledge about the energy consumption or generation capabilities of the microgrid participants, providing a layer of privacy to the participants.

In summation, the way those participants of the microgrid function in the model can be illustrated as a flowchart in Fig. 2.

After modelling the microgrid, modelling of the agent and its action and observation space must be formally defined.

As we've briefly touched on previously, privacy is integrated into the design of the machine learning algorithm. As a result the only data the agent sees from the environment, that is to say from the microgrid, is the power requests, supply offers and the category descriptions of the individual participants.

Power requests and offers are collected individually for each participant. Considering the sample microgrid consists of 3 participants, the observation space would be a 1×3 matrix. Since the energy balances of the participants are unbounded, this observation space can be summarized thusly where the B is the power balance at any given time t.

$$[B_1, B_2, B_3] \quad \forall \quad B \in \mathbb{R} \tag{6}$$

The agent does not take into account just the energy balance of any given hour, but also the categorization of the individual participants of the grid. Observation space for it is as follows. The C in this case represents the category of the participant.

$$[C_1, C_2, C_3] \quad \forall \quad C \in [1, 2, 3] \tag{7}$$

where 1, 2, 3 are the same codes for the participant categorizations in Table 1. Taking all of these into account the observation space that the agent observes is a 1×6 matrix with the previously mentioned characters.

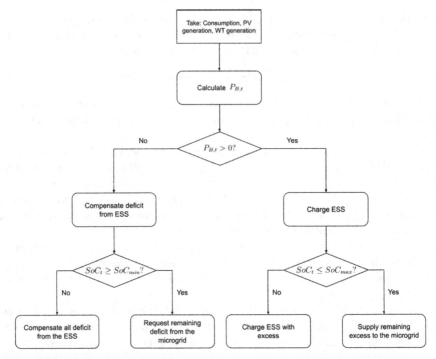

Fig. 2. Flowchart of microgrid consumer activities.

Once the observation space is established, the agent will need to act according to the knowledge it receives from the observation space. The actions the agent can take in this context can be divided into two.

First action the agent can take is to divide the internally offered energy to the microgrid participants that may need the energy, thereby using the excess internally. This action is represented as a 1×4 matrix. The first three columns are for the amount of power the participants receive from the internal surplus. The fourth column is for the total energy exported outside of the grid for reward. This reward is assigned to teach the agent that any excess energy exported outside the grid will have a monetary benefit to the microgrid, hence it's being rewarded. This matrix could be summed up as follows and I is shorthand for internally used power and X is shorthand for exported power.

$$[I_1, I_2, I_3, X] \quad \forall \quad I, X \in \mathbb{R} \tag{8}$$

For the cases where the internal surplus energy is not enough, externally supplied energy is needed, for these cases the agent can assign how much power each of the participants will have from outside the grid at each time step t. The imported power M is added to the action space as a 1×3 matrix thusly,

$$[M_1, M_2, M_3] \quad \forall \quad M \in \mathbb{R} \tag{9}$$

Once the appropriate power is supplied to the grid through the action space, the actions will be judged by the environment and rewards will be assigned.

A well-designed reward system, as was explained in [20], is of utmost importance for the functioning of the machine learning model. The rewards and punishments are used to guide the agent to a policy, that is to say internal to logic, that would be in line with how a microgrid should be run. To do this, first rewards should be assigned to successful usage of internal excess energy within the grid. If there is still excess energy within the microgrid, then the rest should be exported. Lastly, in the case of energy shortages within the microgrid, external sources should be utilized to balance the deficit. To achieve this goal the following rewards are assigned (Table 2).

Table 2. Rewards for the actions.

Participant category	Categorization code
Critical	1
Vulnerable	2
Normal	3

This reward scheme alone is not enough to support the activity of the microgrid, as there is no punishment for simply not supplying the participants in case of an energy deficit within the microgrid. To make sure the agent will import energy from outside despite the negative reward, i.e. punishment, specific negative rewards are assigned to each category of microgrid participants. This auxiliary reward scheme could be summed up as follows (Table 3).

Table 3. Negative rewards for power shortages in any given time for one hour.

Participant category	Categorization code
Critical	1
Vulnerable	2
Normal	3

With these configurations, the agent is set to explore and learn from the environment to manage the microgrid.

5   Results

After taking 15 million actions and receiving rewards from the environment, the agent is introduced to a different data set of consumer data. The process of learning with 15 million cases takes around a day while returning with a course of action for the test data is almost instantaneous. The results are recorded for 48 consecutive hours for one sample episode. The results for the entire microgrid could be observed in Fig. 3.

The first thing to note in the figure is that the model prefers to use a mixture of internal and externally sourced energy for most of the episode. This may be due to a few reasons. The most likely explanation for this behavior could be that the model tries to avoid not being able to supply an adequate amount of energy to the participant. To achieve this goal the model purposefully takes a less than optimal route to make sure it does not incur the penalty of not being able to supply sufficient amounts of energy to the system.

Similarly, after the tenth hour of the episode, we begin to observe energy exports. It is possible that the model has recognized the daily pattern of energy generation and consumption and decided that it is the optimal time slot to begin exporting the excess energy that it possesses in order to maximize the amount of reward earned in the episode. It is worthwhile to note that the model continued importing energy into the microgrid while exporting at the same time. It is possible that the agent has chosen this course of action due to the fact that there is more reward in exporting energy than there is punishment for importing, thereby it seems logical to the model to export the energy at hand and import a token amount to make sure that there is no energy deficit within the grid.

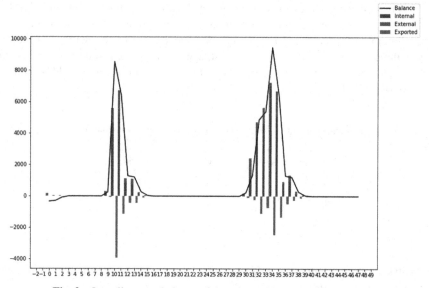

Fig. 3. Overall energy balance of the microgrid. (Color figure online)

A similar picture also exists for the critical participant of the microgrid, a great excess is consistently maintained by the grid controller, presumably to make sure there is no shortage in the system. There are moments where not all of the supply is used neither by the internal nor by the external consumption, causing a space between line and bars. This is due to the fact that excess energy is exported to maximize the reward of the microgrid. Despite the export of energy, we also observe sharing other participants' production, shown as green bars. Adding to that some amount of energy is imported from outside the microgrid to make sure there is no shortage in the microgrid.

Furthermore, throughout the first day there is no supply and demand, pointing to the fact that excesses in that time are used to charge the energy storage systems within the grid. The next morning the excesses left after the energy storage systems are fully charged some amount of energy is coming from the grid controller and the excess is exported (Fig. 4).

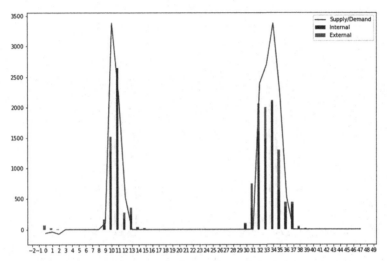

Fig. 4. Overall energy balance in the critical participant of the microgrid. (Color figure online)

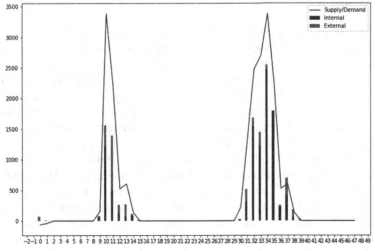

Fig. 5. Overall energy balance in the vulnerable participant of the microgrid. (Color figure online)

A similar picture is also seen with the vulnerable participant of the microgrid. In Graph 5 it is observed that a greater liberty is taken with the excess amount, as seen by the larger gap between the supply line and the supplied amount seen as the bar. The difference in between is sold as export to the grid.

Mismatches can be observed in the activities when it comes to the normal participant of the microgrid. While the agent is mostly successful in making sure there is no deficit in the grid, there exists cases where the supply and demand are mismatching, as can be seen in Graph 6.

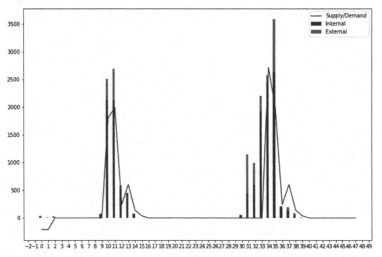

Fig. 6. Overall energy balance in the normal participant of the microgrid. (Color figure online)

6 Conclusions

Climate change caused by global warming is an ever-increasing problem in the global community. One method at our disposal in overcoming this problem is the increased usage of renewable energy sources in the energy mix. To spare the utility grid of the complications of the more intermittent nature of renewable energy generation, this paper takes into account a sample microgrid where these energy sources would be utilized. The model proposed in this paper attempts to automate the operations within this microgrid with regards to energy dispatch, as well as the energy imports and exports.

To that end, reinforcement learning is chosen due to its reward-based structure and ability to make decisions based on incomplete knowledge. Reinforcement learning algorithm chosen for this application is the proximal policy optimization. The reason for this choice is the fact that it's newer than the other algorithms that have been covered in this paper, such as Q-Learning. The PPO also has the distinction of being a policy optimization algorithm which means that the model, rather than maximizing directly for the reward earned from each move, optimizes itself for the probability of earning the most reward at the end of the episode, without falling to the common pitfalls faced by other policy optimization algorithms such as low sample efficiency and performance issues.

A simulation of the microgrid system is built specifically for the sample microgrid using OpenAI Gym. Here lies one of the novelties of this paper, the RL learning environment constructed for this application does not convey the information of energy generation and consumption directly to the agent. The only information received by the agent is the supply and demand of the participants and their corresponding criticality. This criticality information depends on the nature of the participant and ranges from critical to normal depending on whether any energy losses are acceptable considering their importance to the grid, and to the fact whether they are at risk of not being able to supply their own energy. Such categorization, to the best of the author's knowledge, does not exist in other RL applications in the field of microgrid energy dispatch.

The agent is set to perform 15M timesteps in different environmental conditions with a variety of energy consumption profiles to learn how to manage the energy dispatch within the microgrid. The performance of the system displays that the proximal policy optimization algorithms can feasibly be used in the governance of a microgrid. While it is possible to observe cases where participants were importing energy while there is an energy surplus, this can potentially be attributed to the uncertainty of the environment.

This is the first work to feature PPO and participant criticality within the microgrid. This work could be further in the future with the improvement reward system to include stricter punishments for the oversupply of the energy, implementation of a carbon tax to further prioritize renewable energy sources and the usage of publicly owned ESS and renewable energy sources, which could be used either to extract revenue from the system, or to further reduce the energy dependency of the microgrid.

References

1. Sahin, A.: Progress and recent trends in wind energy. Prog. Energy Combust. Sci. **30**(5), 501–543 (2004)
2. Duman, A.C., Güler, Ö.: Economic analysis of grid-connected residential rooftop PV systems in Turkey. Renew. Energy **148**, 697–711 (2020)
3. Rezaeiha, A., Montazeri, H., Blocken, B.: A framework for preliminary large-scale urban wind energy potential assessment: roof-mounted wind turbines. Energy Convers. Manag. **214**, 112770 (2020)
4. Vineetha, C.P., Babu, C.A.: Smart grid challenges, issues and solutions. In: 2014 International Conference on Intelligent Green Building and Smart Grid (IGBSG), Taipei, Taiwan, pp. 1–4 (2014)
5. Majzoobi, A., Khodaei, A.: Application of microgrids in supporting distribution grid flexibility. IEEE Trans. Power Syst. **32**(5), 3660–3669 (2017)
6. Guerrero, J.M., Vasquez, J.C., Matas, J., de Vicuna, L.G., Castilla, M.: Hierarchical control of droop-controlled AC and DC microgrids—a general approach toward standardization. IEEE Trans. Ind. Electron. **58**(1), 158–172 (2011)
7. Farrokhabadi, M., Canizares, C.A., Simpson-Porco, J.W., Nasr, E., Fan, L., Mendoza-Araya, P.A., et al.: Microgrid stability definitions, analysis, and examples. IEEE Trans. Power Syst. **35**(1), 13–29 (2020)
8. Wu, D., Zheng, X., Xu, Y., Olsen, D., Xia, B., Singh, C., et al.: An open-source extendable model and corrective measure assessment of the 2021 texas power outage. Adv. Appl. Energy **4**, 100056 (2021)

9. Murty, V.V.S.N., Kumar, A.: Multi-objective energy management in microgrids with hybrid energy sources and battery energy storage systems. Prot. Control Mod. Power Syst. **5**(1) (2020). Article number: 2. https://doi.org/10.1186/s41601-019-0147-z
10. Qazi, H.S., Liu, N., Wang, T.: Coordinated energy and reserve sharing of isolated microgrid cluster using deep reinforcement learning. In: 2020 5th Asia Conference on Power and Electrical Engineering (ACPEE), Chengdu, China, pp. 81–86 (2020)
11. Kozlov, A.N., Tomin, N.V., Sidorov, D.N., Lora, E.E.S., Kurbatsky, V.G.: Optimal operation control of PV-biomass gasifier-diesel-hybrid systems using reinforcement learning techniques. Energies **13**(10), 2632 (2020)
12. Muriithi, G., Chowdhury, S.: Optimal energy management of a grid-tied solar PV-battery microgrid: a reinforcement learning approach. Energies **14**(9), 2700 (2021)
13. Shang, Y., et al.: Stochastic dispatch of energy storage in microgrids: An augmented reinforcement learning approach. Appl. Energy **261**, 114423 (2020)
14. Al-Gabalawy, M.: Advanced machine learning tools based on energy management and economic performance analysis of a microgrid connected to the utility grid. Int. J. Energy Res., 1–22 (2021). https://doi.org/10.1002/er.6764
15. Sadeghi, M., Mollahasani, S., Erol-Kantarci, M.: Power loss-aware transactive microgrid coalitions under uncertainty. Energies **13**(21), 5782 (2020)
16. Fan, L., Zhang, J., He, Y., Liu, Y., Hu, T., Zhang, H.: Optimal scheduling of microgrid based on deep deterministic policy gradient and transfer learning. Energies **14**(3), 584 (2021)
17. Samadi, E., Badri, A., Ebrahimpour, R.: Decentralized multi-agent based energy management of microgrid using reinforcement learning. Int. J. Electr. Power Energy Syst. **122**, 106211 (2020)
18. Fang, X., Zhao, Q., Wang, J., Han, Y., Li, Y.: Multi-agent deep reinforcement learning for distributed energy management and strategy optimization of microgrid market. Sustain. Cities Soc. **74**, 103163 (2021)
19. Guo, C., Wang, X., Zheng, Y., Zhang, F.: Optimal energy management of multi-microgrids connected to distribution system based on deep reinforcement learning. Int. J. Electr. Power Energy Syst. **131**, 107048 (2021)
20. Wu, T., Wang, J.: Artificial intelligence for operation and control: the case of microgrids. Electricity J. **34**(1), 106890 (2021)
21. Yin, L., Zhang, B.: Time series generative adversarial network controller for long-term smart generation control of microgrids. Appl. Energy **281**, 116069 (2021)
22. Sutton, R.S., Barto, A.G.: Reinforcement Learning: An Introduction, 2nd edn. MIT Press, Cambridge (2018)
23. Gu, S., Lillicrap, T., Ghahramani, Z., Turner, R.E., Levine, S.: Q-Prop: sample-efficient policy gradient with an off-policy critic. ArXiv (2017)
24. Open AI Proximal Policy Optimization. https://openai.com/blog/openai-baselines-ppo. Accessed 09 Feb 2021
25. Introduction to Deep Reinforcement Learning (Deep RL). https://www.youtube.com/watch?v=zR11FLZ-O9M&t=3487s. Accessed 09 Feb 2021
26. Schulman, J., Wolski, F., Dhariwal, P., Radford, A., Klimov, O.: Proximal policy optimization algorithms. arXiv (2017)
27. EU Energy Poverty Observatory, What is energy poverty? https://www.energypoverty.eu/about/what-energy-poverty. Accessed 09 Jan 2021
28. Photovoltaic Geographical Information System (PVGIS). https://ec.europa.eu/jrc/en/pvgis. Accessed 09 Feb 2021

29. Atia, R., Yamada, N.: Sizing and analysis of renewable energy and battery systems in residential microgrids. IEEE Trans. Smart Grid **7**(3), 1204–1213 (2016)
30. Kim, R.-K., Glick, M.B., Olson, K.R., Kim, Y.-S.: MILP-PSO combined optimization algorithm for an islanded microgrid scheduling with detailed battery ESS efficiency model and policy considerations. Energies **13**(8), 1898 (2020)

Residential Short-Term Load Forecasting via Meta Learning and Domain Augmentation

Di Wu[1(✉)], Can Cui[2], and Benoit Boulet[1]

[1] McGill University, Montreal, QC H3A 0E9, Canada
di.wu5@mail.mcgill.ca, benoit.boulet@mcgill.ca
[2] Google LLC, New York, NY 10044, USA
tracycui@google.com

Abstract. With the increasing adoption of electric devices and renewable energy generation, electric load forecasting, especially short-term load forecasting (STLF), has recently attracted more attention. Accurate short-term load forecasting is of significant importance for the safe and efficient operation of power grids. Deep learning-based models have achieved impressive success on several applications, including short-term load forecasting. Yet, most deep learning models do require a large amount of training data. However, in the real world, it may be very difficult or even impossible to collect enough data to train a reliable machine learning model. This makes is hard to adopt deep models for several real-world scenarios. Thus, it will be very helpful if deep learning models can be learned to tackle tasks with limited amount of training data and unseen tasks. In this work, we propose to use the meta-learning framework to train a long short-term memory-based model for short-term residential load forecasting. Specifically, by minimizing the task-level loss (loss over several tasks), the model is trained to perform well on different tasks. We also use domain randomization techniques to further augment the training tasks, which may further improve the generalization ability of the proposed model. Our model is evaluated on real-world data sets and compared against some classic forecasting models.

Keywords: Electric load forecasting · Meta learning

1 Introduction

Electric load forecasting aims to forecast the future load consumption based on a set of historical records as well as other external factors [1]. It has been shown that accurate electric load forecasting is of significant importance for the efficient operation of modern power grids. The forecasting accuracy can help significantly reduce the total operation cost [2,3]. As shown [4] in 1% increase of forecasting error would cause a £10 million increase in operating cost per year for the UK power systems. Depending on the forecasting horizon, electric

© IFIP International Federation for Information Processing 2022
Published by Springer Nature Switzerland AG 2022
E. Mercier-Laurent and G. Kayakutlu (Eds.): AI4KMES 2021, IFIP AICT 637, pp. 184–196, 2022.
https://doi.org/10.1007/978-3-030-96592-1_14

load forecasting ranges from short-term forecasting (minutes or hours ahead) to long-term forecasting (years ahead) [5]. Short-term forecasting, main focus of this work, is mainly used to support real-time energy dispatch.

Electric load forecasting could be very challenging due to multiple uncertain factors. Specifically, there are uncertain factors for both the energy demand side and the energy generation side. For example, in recent years, different types of electrical appliances are integrated into the power grids [2], such as electric vehicles [6–12]. With the high penetration increase of electric vehicles, there will be a corresponding charging consumption. This electric consumption will be highly affected by human behaviors. On the other side the renewable energy generation has also been increasing very quickly in recent years. As shown in [13], the renewable energy generation has been increasing almost exponentially in the past ten years. Due to these reasons, a single residential unit may be more challenging than an industrial building. This can be more challenging when only limited data is available. In our work, we mainly focus on tackling the electric load forecasting for single houses.

Electric load forecasting has been an important research topic for the past few years. In general, the electric load forecasting algorithms can be generally categorized into two groups: statistical [14–16] and machine learning methods [5,17–22]. The benefit of autoregressive integrated moving average (ARIMA) model was showcased in [15,16]. Different types of machine learning-based methods have also been utilized for electric load forecasting including support vector regression (SVR) [18,19], general additive models [21], neural networks (NNs) [23,24], and random forest [25,26].

Most of the current works assume that we have a large amount of training data and stable data distribution [27]. In the real world, we may need to deal with a new house with a limited amount of data or even no labeling data. However, most of the current machine learning-based models would require a large amount of data to learn a reliable forecasting model. Without enough data, there will be a significant performance degradation for the model performance. Thus it will be very beneficial if we can learn a robust forecasting model that is robust to the changes of forecasting tasks. In this work, we propose to use meta-learning [28–32] to deal with this challenge. Specifically, we mains to use meta learning and data augmentation to learn a good model initialization based on which the model can adapt fast on a new task.

The remainder of this paper is organized as follows. The technical background of this paper is presented in Sect. 2. The short-term load forecasting method is presented in Sect. 3. Real-world data sets based evaluation results are presented in Sect. 4. Finally, the conclusion and future work are presented in Sect. 5.

2 Background

2.1 Meta Learning

Deep learning models have achieved impressive success for different types of applications in recent years. However, most of these applications would require

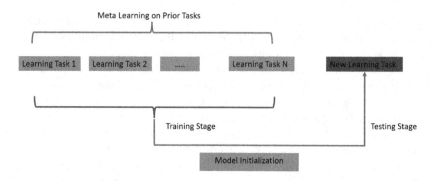

Fig. 1. Meta learning concept

a huge amount of training data which can be very challenging in real world. Meta-learning [33,34], also known as learning to learn, aims to empower the model to learn fast with a limited amount of data. This can be very helpful for machine learning models, especially for deep learning based models which usually would require a large amount of training data. As shown in Fig. 1, different from other classic learning paradigms, meta learning focuses on improving the task level generalization. After finishing the learning on multiple tasks, the meta learner can perform better, i.e., perform better after experiencing the same number of samples from the new task.

Different from traditional machine learning paradigms, the meta-learner is trained on a distribution of similar tasks, with the goal of learning a strategy can generalize to related but unseen tasks from a similar task distribution. Depending on how the meta learner is trained, there mainly three types of meta learning method, i.e., metric based methods [35–37], optimization based methods [28], and context based methods [38]. Optimization based method aims to learn a good initialization which can enable the meta learner to perform fast adaptation once presented with a new task.

2.2   Time Series Forecasting

The short-term load forecasting can be treated as a time series problem. Time series forecasting aims to predict the future values of time series data given the observed history. Different types of features have shown to be effective for electric load forecasting. In our work, we aims to utilize the three types of features including lagged electric load, lagged temperature information, and weekday/weekend information for the one hour ahead electric load forecasting. One-hour ahead electric load forecasting can be very helpful for the real-time energy dispatching.

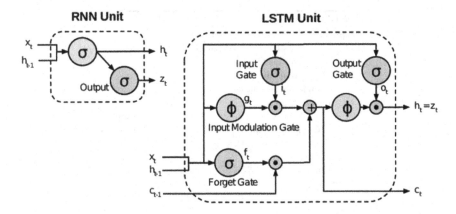

Fig. 2. LSTM unit and RNN unit

2.3   LSTM Based Load Forecasting

Neural networks (NN), especially deep neural networks have shown impressive breakthroughs for different applications due to their strong representation power [39,40]. Typical neural networks include feed forward neural networks, convolutional neural networks, and recurrent neural networks (RNNs). Feed forward neural network is the most basic type of NNs which consists of an input layer, one or more fully connected hidden layers, and a final output layer. Each hidden layer consists of several neurons realizing the non-linear transformation of outputs from previous layer.

Different from feed forward neural networks, RRNs are sequence based models and the information from previous timesteps can be used for the current timestep. This makes RNNs very suitable to model temporal correlated data and deal with time series problems. RNNs are mainly trained with backpropagation through time. However, there could be gradient vanishing and gradient exploding problems when training long-range dependencies with vanilla RNNs models [41]. The comparison of RNN unit and LSTM unit is presented in Fig. 2[1]. As shown in this Figure, we can see that there are multiple additional cells for LSTM. Due to its excellent ability on sequential modelling, in this paper, LSTM based model is adopted for short-term residential load forecasting.

3   Methodology

Most of the current forecasting method would train a forecasting based on a fixed historical data sets. This may limit the generalization ability for In this paper, we propose to utilize meta learning to improve the model generalization for the

---

[1] Figure is from [42].

Algorithm 1 Model-Agnostic Meta-Learning

Require: $p(\mathcal{T})$: distribution over tasks
Require: α, β: step size hyperparameters
1: randomly initialize θ
2: **while** not done **do**
3:    Sample batch of tasks $\mathcal{T}_i \sim p(\mathcal{T})$
4:    **for all** $\mathcal{T}_i$ **do**
5:       Evaluate $\nabla_\theta \mathcal{L}_{\mathcal{T}_i}(f_\theta)$ with respect to K examples
6:       Compute adapted parameters with gradient descent: $\theta_i' = \theta - \alpha \nabla_\theta \mathcal{L}_{\mathcal{T}_i}(f_\theta)$
7:    **end for**
8:    Update $\theta \leftarrow \theta - \beta \nabla_\theta \sum_{\mathcal{T}_i \sim p(\mathcal{T})} \mathcal{L}_{\mathcal{T}_i}(f_{\theta_i'})$
9: **end while**

Fig. 3. Overview for the model agnostic meta learning

electric load forecasting. Figure 4 presents an overview for the main objective of this paper. Our primal objective is to learning a good model initialization that can perform well on load forecasting for a new house.

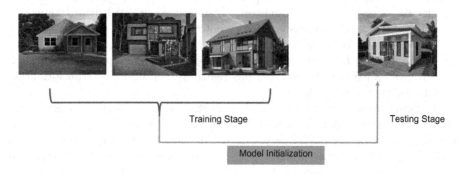

Fig. 4. Application of meta learning for single house load forecasting

3.1 Model Agnostic Meta Learning

MAML is one of gradient based meta learning method and has shown to be effective for different types of applications such as traffic light control [43], image classification [44], and nature language processing [45]. MAML aims to learn a good model initialization based on multiple sampled tasks. As shown in Fig. 3, there are mainly two loops of optimization. In the first loop, a large amount of tasks will be sampled and we will then learn a model for each sampled tasks based on the limited amount of data of that specific task. In the second loop, the model initialization θ^* will be learned based on optimizing the meta loss.

3.2   MAML Based Electric Load Forecasting

In this work, we propose to use a MAML based forecasting framework to learn a robust forecasting model which will be able to deal with different forecasting tasks, i.e., forecasting for different houses including unseen houses. Figure 5 shows the overview of the two-stage forecasting framework. Specifically, in the first stage we will pretrain a base model θ_0 over multiple (N) houses. In this stage, we assume that we have a large amount of the training data for these houses. In this paper, the base model is based on LSTM. It is worth noting that the proposed framework could also be applied for other base forecasting models. The second stage is normal MAML training but with the learned model parameters as initialization θ_0 rather than random initialization. θ_0 will be updated through training iterations.

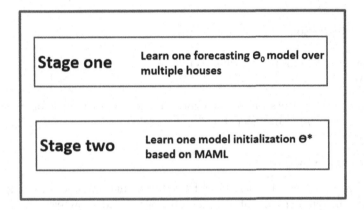

Fig. 5. Overview of the two stage forecasting method

4   Experiment Results

4.1   Dataset

In this paper, the electric load data sets from OpenEI are used to evaluate the effectiveness of proposed forecasting algorithms. In this paper, three types of features including lagged electric load (electricity consumed in the last four hours), lagged temperature information (temperature in the last four hours), and weekday/weekend information (1 for weekday and 0 for the weekend) are used for short-term electric load forecasting. To accelerate the training process, the feature normalization is implemented for all features. The used data set include one-year (2014) hourly load consumption data (8760 data points) for 72 houses in New York. We focus on the load consumption in winter times, particularly in the load consumption of December month.

4.2   Baselines and Evaluation Metrics

In this paper, mean average percentage error (MAPE) as shown in Eq. 1 and mean absolute error (MAE) as shown in Eq. 2 are used to evaluate the effectiveness of the proposed algorithms. As shown in these two equations, y_i' is the predicted load consumption and y_i is the real value for load consumption at time slot i.

$$MAPE = \frac{\sum_{i=1}^{N} \frac{|y_i - y_i'|}{y_i}}{N} \qquad (1)$$

$$MAE = \frac{\sum_{i=1}^{N} |y_i - y_i'|}{N} \qquad (2)$$

For benchmark results, four other frequently used forecasting methods are used as baseline models:

- Linear Regression
- SVR
- Feed-froward Neural Network (NN)
- LSTM

We compare the results of the baselines and proposed model for experiments described in Sect. 4.5 and Sect. 4.6.

4.3   Experiment Setup

As mentioned in Sect. 4.1, the OpenEI dataset contains load consumption for 72 houses, in our experiment we pick 10 houses and treat each household as a class. The December load consumption of three randomly picked classes are shown in Fig. 6, as observed, the load consumption can vary a lot among different households. December load consumption of all 10 houses are combined, we train the models on December load consumption of other households (training classes) to predict the load for a particular household (test class).

4.4   Hyper Parameters

All baseline models are tuned with parameters with the best performance for the data set. Especially: For SVR, we choose RBF kernel, $C = 1e3$, $\epsilon = 0.1$, $\gamma = 0.1$; For NN, we choose a 20 layer Multi-Layer Perceptron (MLP) with Rectified Linear Unit (ReLU) as the activation function, $\alpha = 0.001$, 1000 maximum iterations; For LSTM, we build the LSTM model with an LSTM layer of 64 nodes, a dense layer with 128 nodes and use Adam as the optimizer. The data is broken into batches of 128 and we train the model for 1000 epochs. For MAML, the inner step size is 0.002, the outer step size is 0.005, the sample size is 20 for Sect. 4.5 and Sect. 4.6 and has selected values for Sect. 4.7.

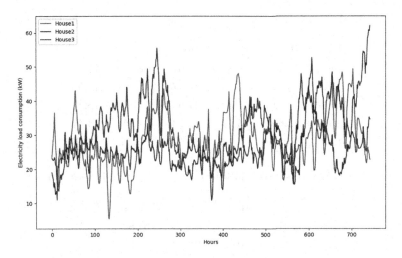

Fig. 6. December load consumption of three houses

Table 1. MAPE (%) for short-term load forecasting on five houses

Method	House 1	House 2	House 3	House 4	House 5
Linear	56.07	60.12	57.89	65.12	95.54
SVR	55.47	58.86	56.91	58.75	94.09
NN	53.52	60.28	59.97	64.95	95.47
LSTM	46.77	50.97	47.62	51.55	81.22
MAML	**24.94**	**23.10**	**24.17**	**19.09**	**27.65**

4.5   Evaluations on Five Houses

We evaluate the performance of the baseline and meta-learning models and reported the results on five houses. When training meta-learning model, we sample batch of tasks of 20 shots from training data then fine-tune and evaluate on test data. Table 1 and Table 2 respectively show the average forecasting MAPE (%) and MAE of different models on the five houses. As shown in the tables, the baseline models all have fairly large MAPE and MAE in general and perform particularly bad on House 5, while the meta-learning model has consistent performance across houses and shows on average a 55.9% improvement for MAPE and a 59.21% improvement for MAE over the best baseline model (LSTM). These results show that the model trained by meta-learning method generalize well to unseen data distributions while baseline models can fail badly when the test and training data has different distributions (House 5 as an example).

Table 2. MAE for short-term load forecasting on five houses

Method	House 1	House 2	House 3	House 4	House 5
Linear	15.13	15.39	17.01	18.12	15.61
SVR	15.11	15.26	16.77	16.57	15.55
NN	14.11	15.32	17.63	17.95	15.48
LSTM	12.60	13.21	14.23	14.55	13.44
MAML	**5.95**	**5.66**	**6.75**	**5.37**	**3.97**

Table 3. MAPE (%) for short-term load forecasting on five houses with noisy data

Method	House 1	House 2	House 3	House 4	House 5
Linear	48.15	52.04	51.61	52.64	83.34
SVR	45.18	50.18	48.94	46.03	81.63
NN	48.45	52.27	51.83	51.54	83.74
LSTM	42.00	48.63	45.88	44.82	79.76
MAML	**23.93**	**23.18**	**23.76**	**19.14**	**26.64**

4.6 Evaluations on Five Houses with Noisy Data

To further showcase the robustness of the proposed method, we compare its performance with other baselines on noisy data. Specifically, we add Gaussian noise of mean 0 and standard deviation 0.5 to all data. Table 3 and Table 4 respectively show the average forecasting MAPE (%) and MAE of different models on the five houses with noisy data. As shown in the tables, MAML still consistently has the best performance across houses and shows on average a 53.96% improvement for MAPE and a 56.07% improvement for MAE over the best baseline model (LSTM). Compared to Table 1 and Table 2, the performance of baseline models has improved because the Gaussian noise reduced the effect of training datas' distribution on the testing data, thus improved the model's ability to generalize to new data.

4.7 MAML Training with Different Number of Shots

To evaluate how MAML performs when the number of selected samples (shots) in training step varies, we train MAML with different number of shots, more specifically 1-shot, 5-shot, 10-shot, 20-shot, 30-shot and 50-shot, and report the results in Table 5 and Table 6. As shown in the results, the performance remains steady across 1/5/10/20/30/50-shot training for all houses.

Table 4. MAE for short-term load forecasting on five houses with noisy data

Method	House 1	House 2	House 3	House 4	House 5
Linear	12.83	13.19	15.20	14.97	13.53
SVR	12.05	12.91	14.42	13.35	13.26
NN	12.81	13.21	15.20	14.70	13.60
LSTM	11.27	12.46	13.50	13.01	13.01
MAML	**5.90**	**5.63**	**6.63**	**5.40**	**4.10**

Table 5. MAPE (%) for short-term load forecasting on five houses with 1/5/10/20/30/50-shot MAML training

House	1-shot	5-shot	10-shot	20-shot	30-shot	50-shot
House 1	22.85	22.98	23.04	22.71	22.64	23.19
House 2	21.84	21.94	22.51	23.08	22.73	23.24
House 3	22.14	22.68	23.22	23.11	24.23	23.31
House 4	17.83	18.74	18.49	17.81	19.13	18.70
House 5	26.04	25.78	26.43	25.81	26.84	26.06

Table 6. MAE for short-term load forecasting on five houses with short-term load forecasting on five houses with 1/5/10/20/30/50-shot MAML training

House	1-shot	5-shot	10-shot	20-shot	30-shot	50-shot
House 1	5.64	5.67	5.60	5.67	5.69	5.78
House 2	5.29	5.36	5.40	5.37	5.46	5.50
House 3	6.22	6.26	6.37	6.38	6.50	6.44
House 4	5.10	5.23	5.23	5.04	5.30	5.29
House 5	3.90	4.00	3.88	3.89	3.96	3.98

5 Conclusion

Accurate electric load forecasting is of significant importance for the safe and economic operation of power grids. Most current machine learning-based forecasting models assume that we have enough training data and that new data would follow similar distribution as training data. However, these assumptions may not be true for real-world applications. Thus it will be very beneficial if we can train a model with existing data that is robust and adapts quickly to new incoming data. In this work, we propose to use the meta-learning framework to learn a robust machine learning model for residential load forecasting and ran prototyping experiments to evaluate its performance. Experiment results are implemented on real-world data sets to showcase the effectiveness of the proposed method. For future work, we plan to further improve the proposed framework and design more comprehensive experiments for evaluation.

References

1. Hahn, H., Meyer-Nieberg, S., Pickl, S.: Electric load forecasting methods: tools for decision making. Eur. J. Oper. Res. **199**(3), 902–907 (2009)
2. Wu, D.: Machine Learning Algorithms and Applications for Sustainable Smart Grid. McGill University (Canada) (2018)
3. Wu, D., Zeng, H., Lu, C., Boulet, B.: Two-stage energy management for office buildings with workplace EV charging and renewable energy. IEEE Trans. Transp. Electrification **3**(1), 225–237 (2017)
4. Bunn, D., Farmer, E.D.: Comparative models for electrical load forecasting (1985)
5. Hippert, H.S., Pedreira, C.E., Souza, R.C.: Neural networks for short-term load forecasting: a review and evaluation. IEEE Trans. Power Syst. **16**(1), 44–55 (2001)
6. Hall, D., Lutsey, N.: Electric vehicle charging guide for cities. Consulting Report (2020)
7. Wu, D., Zeng, H., Boulet, B.: Neighborhood level network aware electric vehicle charging management with mixed control strategy. In: 2014 IEEE International Electric Vehicle Conference (IEVC), pp. 1–7. IEEE (2014)
8. Wu, D., Rabusseau, G., François-lavet, V., Precup, D., Boulet, B.: Optimizing home energy management and electric vehicle charging with reinforcement learning. In: Proceedings of the 16th Adaptive Learning Agents (2018)
9. Dang, Q., Wu, D., Boulet, B.: EV charging management with ANN-based electricity price forecasting. In: 2020 IEEE Transportation Electrification Conference and Expo (ITEC), pp. 626–630. IEEE (2020)
10. Dang, Q., Wu, D., Boulet, B.: An advanced framework for electric vehicles interaction with distribution grids based on q-learning. In: 2019 IEEE Energy Conversion Congress and Exposition (ECCE), pp. 3491–3495. IEEE (2019)
11. Dang, Q., Wu, D., Boulet, B.: A q-learning based charging scheduling scheme for electric vehicles. In: 2019 IEEE Transportation Electrification Conference and Expo (ITEC), pp. 1–5. IEEE (2019)
12. Dang, Q., Wu, D., Boulet, B.: EV fleet batteries as distributed energy resources considering dynamic electricity pricing. In: 2021 IEEE 12th International Symposium on Power Electronics for Distributed Generation Systems (PEDG), pp. 1–6. IEEE (2021)
13. Adebayo, T.S., et al.: Modeling the dynamic linkage between renewable energy consumption, globalization, and environmental degradation in South Korea: does technological innovation matter? Energies **14**(14), 4265 (2021)
14. Contreras, J., Espinola, R., Nogales, F.J., Conejo, A.J.: Arima models to predict next-day electricity prices. IEEE Trans. Power Syst. **18**(3), 1014–1020 (2003)
15. He, H., Liu, T., Chen, R., Xiao, Y., Yang, J.: High frequency short-term demand forecasting model for distribution power grid based on ARIMA. In: IEEE CSAE, vol. 3, (Zhangjiaji, China), pp. 293–297 (2012)
16. Matsila, H., Bokoro, P.: Load forecasting using statistical time series model in a medium voltage distribution network. In: IEEE IECON, (Washington, DC), pp. 4974–4979 (2018)
17. Wu, D., Wang, B., Precup, D., Boulet, B.: Multiple kernel learning-based transfer regression for electric load forecasting. IEEE Trans. Smart Grid **11**(2), 1183–1192 (2019)
18. Ye, J., Yang, L.: A comparative study of ensemble support vector regression methods for short-term load forecasting. In: IEEE ICSAI, (Nanjing, China), pp. 139–143 (2018)

19. Chen, Y., et al.: Short-term electrical load forecasting using the support vector regression (SVR) model to calculate the demand response baseline for office buildings. Appl. Energy **195**, 659–670 (2017)
20. Dong, Y., Zhang, Z., Hong, W.-C.: A hybrid seasonal mechanism with a chaotic cuckoo search algorithm with a support vector regression model for electric load forecasting. Energies **11**(4), 1009 (2018)
21. Wu, D., Wang, B., Precup, D., Boulet, B.: Boosting based multiple kernel learning and transfer regression for electricity load forecasting. In: Altun, Y., et al. (eds.) ECML PKDD 2017. LNCS (LNAI), vol. 10536, pp. 39–51. Springer, Cham (2017). https://doi.org/10.1007/978-3-319-71273-4_4
22. Lin, W., Wu, D., Boulet, B.: Spatial-temporal residential short-term load forecasting via graph neural networks. IEEE Trans. Smart Grid **12**(6), 5373–5384 (2021)
23. Kong, W., Dong, Z.Y., Jia, Y., Hill, D.J., Xu, Y., Zhang, Y.: Short-term residential load forecasting based on LSTM recurrent neural network. IEEE Trans. Smart Grid **10**(1), 841–851 (2019)
24. Kim, N., Kim, M., Choi, J.K.: LSTM based short-term electricity consumption forecast with daily load profile sequences. In: IEEE GCCE, Las Vegas, pp. 136–137 (2018)
25. Moon, J., Kim, Y., Son, M., Hwang, E.: Hybrid short-term load forecasting scheme using random forest and multilayer perceptron. Energies **11**(12), 3283 (2018)
26. Son, M., Moon, J., Jung, S., Hwang, E.: A short-term load forecasting scheme based on auto-encoder and random forest. In: Ntalianis, K., Vachtsevanos, G., Borne, P., Croitoru, A. (eds.) APSAC 2018. LNEE, vol. 574, pp. 138–144. Springer, Cham (2019). https://doi.org/10.1007/978-3-030-21507-1_21
27. Hoens, T.R., Polikar, R., Chawla, N.V.: Learning from streaming data with concept drift and imbalance: an overview. Progress Artif. Intell. **1**(1), 89–101 (2012)
28. Finn, C., Abbeel, P., Levine, S.: Model-agnostic meta-learning for fast adaptation of deep networks. In: International Conference on Machine Learning, pp. 1126–1135. PMLR (2017)
29. Rusu, A.A., et al.: Meta-learning with latent embedding optimization. arXiv preprint arXiv:1807.05960 (2018)
30. Zintgraf, L., Shiarli, K., Kurin, V., Hofmann, K., Whiteson, S.: Fast context adaptation via meta-learning. In: International Conference on Machine Learning, pp. 7693–7702. PMLR (2019)
31. Olier, I., et al.: Meta-QSAR: a large-scale application of meta-learning to drug design and discovery. Mach. Learn. **107**(1), 285–311 (2018)
32. Mireshghallah, F., Shrivastava, V., Shokouhi, M., Berg-Kirkpatrick, T., Sim, R., Dimitriadis, D.: UserIdentifier: implicit user representations for simple and effective personalized sentiment analysis. arXiv preprint arXiv:2110.00135 (2021)
33. Hospedales, T., Antoniou, A., Micaelli, P., Storkey, A.: Meta-learning in neural networks: a survey. arXiv preprint arXiv:2004.05439 (2020)
34. Giraud-Carrier, C., Vilalta, R., Brazdil, P.: Introduction to the special issue on meta-learning. Mach. Learn. **54**(3), 187–193 (2004)
35. Sung, F., Yang, Y., Zhang, L., Xiang, T., Torr, P.H., Hospedales, T.M.: Learning to compare: Relation network for few-shot learning. In: Proceedings of the IEEE Conference on Computer Vision and Pattern Recognition, pp. 1199–1208 (2018)
36. Snell, J., Swersky, K., Zemel, R.S.: Prototypical networks for few-shot learning. arXiv preprint arXiv:1703.05175 (2017)
37. Santoro, A., Bartunov, S., Botvinick, M., Wierstra, D., Lillicrap, T.: One-shot learning with memory-augmented neural networks. arXiv preprint arXiv:1605.06065 (2016)

38. Santoro, A., Bartunov, S., Botvinick, M., Wierstra, D., Lillicrap, T.: Meta-learning with memory-augmented neural networks. In: International Conference on Machine Learning, pp. 1842–1850. PMLR (2016)
39. He, K., Zhang, X., Ren, S., Sun, J.: Deep residual learning for image recognition. In: Proceedings of the IEEE Conference on Computer Vision and Pattern Recognition, pp. 770–778 (2016)
40. Lillicrap, T.P., et al.: Continuous control with deep reinforcement learning. arXiv preprint arXiv:1509.02971 (2015)
41. Bengio, Y., Simard, P., Frasconi, P., et al.: Learning long-term dependencies with gradient descent is difficult. IEEE Trans. Neural Networks 5(2), 157–166 (1994)
42. Rassem, A., El-Beltagy, M., Saleh, M.: Cross-country skiing gears classification using deep learning. arXiv preprint arXiv:1706.08924 (2017)
43. Zang, X., Yao, H., Zheng, G., Xu, N., Xu, K., Li, Z.: MetaLight: value-based meta-reinforcement learning for traffic signal control. In: Proceedings of the AAAI Conference on Artificial Intelligence, vol. 34, pp. 1153–1160 (2020)
44. Khodadadeh, S., Bölöni, L., Shah, M.: Unsupervised meta-learning for few-shot image classification. arXiv preprint arXiv:1811.11819 (2018)
45. Yin, W.: Meta-learning for few-shot natural language processing: a survey. arXiv preprint arXiv:2007.09604 (2020)

Renewable Energy Investment Decision Evaluation for Local Authorities

Ecem Turkoglu[1]([✉]), Uner Colak[1], Gulgun Kayakutlu[1], and Irem Duzdar Argun[2]

[1] Energy Institute, Istanbul Technical University, Istanbul, Turkey
[2] Engineering Faculty, Duzce University, Duzce, Turkey

Abstract. Currently, increasing urban population creates escalating problems such as housing, infrastructure, transportation, health, environment, safety, and energy consumption. Climate change, emission mitigation, and limited energy supply force urban managers to consider sound measures with the support of technological developments. It is based on data collection and accumulation using IoT, sensors, digital networks, and other means. "Smart urbanism" is a concept that considers predicting, designing, and creating solutions in a systematic, sustainable, and agile manner based on the data collected. Energy is an indispensable dimension in this context. Optimum energy management makes it "smart energy" with the inclusion clean and sustainable renewable energy resources as well as energy efficiency. This study is performed to analyze possible inclusion of geothermal, solar, and wind power. The analyses are based to reveal the best feasible alternative considering the parameters of location, climate, space availability, capital and operational expenditures as well as construction, operation, and maintenance. The Fuzzy Analytic Hierarchy Process (Fussy AHP) technique is used to evaluate the ranking. The proposed method uses fuzzy mathematics for solving problems containing uncertainties as well as less quantifiable. This study proposes a methodological framework for the analysis of competitiveness of alternative renewable energy generations in urban environments. The municipality of Balıkesir is chosen for the case study presented in this work.

Keywords: Fuzzy analytic hierarchy process · Wind energy · Solar energy · Geothermal energy · Energy investments · Urban energy

1 Introduction

Population increase and industrialization increased energy demand in the World. In the past, Turkey has been relied on imported fossil fuel, coal, oil, and natural gas due to the lack of domestic resources in order to match the domestic energy demand. However, there is a global trend to replace fossil fuel based energy systems with renewable energy to cope with and mitigate climate change. Municipalities have also initiated programs for the inclusion of renewable energy into urban utilization. Of course, the choice of renewable energy is mainly determined by the location and availability of renewable resources. For instance, solar and wind energy are more focused in Aegean and Mediterranean regions

E. Mercier-Laurent and G. Kayakutlu (Eds.): AI4KMES 2021, IFIP AICT 637, pp. 197–214, 2022.
https://doi.org/10.1007/978-3-030-96592-1_15

whereas biomass is more preferred in Marmara region [1]. Geothermal energy is more site dependent and can be utilized wherever it is available [2]. These resources are mostly concentrated in the western regions of Turkey, apart from this, there are also resources in the Northern Anatolian fault line and the volcanic regions in the east [3].

Renewable energy has its own characteristics. The cost of renewable energy becomes more and more affordable as they are widely implemented all around the World. However, renewable energy is mostly site dependent. Wind energy systems may need routine maintenance. Solar energy systems require significant area for implementation. Both solar and wind energy resources are intermittent and not available continuously. On the other hand, geothermal energy has more availability and resources may be stored for proper generation scheduling [4, 5]. Not only the cost, but also characteristics, i.e., local availability, space and maintenance needs, are required for the consideration of renewable energy resources [6].

This study proposes a methodological framework to accurately evaluate the shared economy based on renewable energy resource alternatives. The model is applied for the case of the Municipality of Balıkesir in Turkey. For this reason, within the scope of this study, it will be analyzed which of the renewable energy sources between solar, wind and geothermal energy is more appropriate to invest for the Balikesir municipality on the Aegean region. In this context, the article proceeds by the literature review, then, methodology, and finally, evaluations by the proposed AHP model. At this point, the Saatty scale is converted Triangular Fuzzy Numbers for the pair wise comparison matrices. Finally, the conclusions refer to the one specific renewable energy alternative that better fits for the use in the Municipality of Balikesir.

2   Literature Survey

The lack of domestic energy resources in Turkey also influences the country's economy significantly. There is also a motivation to use renewable resources in order to reduce environmental impacts. The decrease in the cost of renewables, especially for solar and wind, further supports the action for decision makers' energy transition policies and increase investers' appetite. Local municipalities are also within the scope for such actions considering their services. The use of renewable energy by municipalities in their services is also well received by the public.

There are many previous studies to determine the potential of energy resources in Turkey. Topcu et al. [7] used a framework including technical, economic, environmental, and social point of view in their analyses covering solar, wind, hydro, biomass, coal, natural gas, and nuclear energy. They concluded that renewable energy (solar, wind, and hydro) would be beneficial for Turkey. In another study, Celik and Ozgur [8] discussed the current status of solar energy in Turkey with compared to Europe. Emeksiz and Demirci [9] and Argin et al. [10] studied the wind potential in the coastal regions of Turkey. In a separate study, Melikoglu (2017) pointed out the geothermal energy utilization goal, set for 2023 within the Vision 2023 program, had already been met [11].

There are many studies on energy investments for municipalities. Adam et al. [12] evaluated the energy resource that would be the right decision to invest in for the City Council of Leed, UK. In a similar study, Taminiau and Byrne [13] evaluated the solar energy potential of New York City and offered new proposals for the municipality's solar energy approach. In a study conducted in Turkey; Karaca et al. [14] examined the solar energy potential within the municipality of Konya and provided energy by integrating solar panels on the roofs of the headman's houses in the Selçuklu district. In another study performed by Guven [15], some solutions are offered to meet the energy needs of the building of Bahcelievler Municipality. In addition, the study done by Biberci et al. [16] was on the renewable energy utilization of municipalities in Turkey and they concluded that it should be increased.

Gazhelia and Berghb [17], Arikan and Pine [18], Adam et al. [12], and Guven [13] conducted studies focused on decision making between solar and wind energies. Among these studies, Gazhelia and Berghb [17] used the "Real options valuation" technique to make a decision, but this technique did not provide a variety of approaches. On the other hand, Adam et al. [12] took a decision-making approach based on the "Life cycle cost" in which case social and environmental consequences were also taken into account.

In a study carried out by Yilan et al. [19], it was analyzed which sustainable resources Turkey should focus on for electricity production, and in this study, Multi-Criteria Decision Making (MCDM) is employed to analyze by comparing Total Cost of Living, Cost-Benefit Analysis and MCDM. MCDM offers a broader perspective as it provides an approach from many different aspects (technical, environmental, economic, social). Similarly, an application of this method was performed by Li et al. [20] and calculated the most accurate renewable energy investment for China with MCDM techniques.

The use of MCDM techniques are examined by Pohekar and Ramachandran [21] and they concluded that this technique was very useful for energy planning and found that it was being used more and more frequently in this field.

Some studies can be found confirming this result; Dogrusoy and Serin [22] evaluated the city of Izmir on the basis of districts and calculated its energy potentials using the Analytical Hierarchy Process (AHP), which is a MCDM technique. Likewise, Karatop et al. [23] analyzed most suitable energy investment for Turkey using MCDM techniques including Fuzzy AHP. In a similar study Wang et al. [24] use the same technique for select an energy source for Pakistan.

In this study, it will be evaluated which renewable energy source would be more beneficial for a selected municipality on the Aegean region of Turkey. Balikesir is located at the north-west of Anatolia, the coordinates are 270530 °N and 390390 °E, the height above the sea level is 600 m [25]. The total area is 14,292 km^2 and the population is 215,000, the density of population per km^2 is 15 [26].

In this sense, a decision will be made between the solar energy potential that is left behind in use, the wind energy potential that is frequently used in coastal areas and the geothermal energy which Turkey has a big potential especially in the western side of the country. Within the scope of this decision, economic, environmental and social impacts will be considered. To achieve this, an appropriate MCDM technique will be determined. The decision for method selections explained in detail in the Methodology section. The sources mentioned in the literature review are listed chronologically in Table 1.

Table 1. Source table for literature survey

Author	Article	Purpose	MCDM method	Contribution
Pohekar and Ramachandran [21]	Application of multi-criteria decision making to sustainable energy planning—A review	The applications of MCDM in energy planning have been examined	PROMETHEE (Preference ranking organization method for enrichment evaluation) and ELECTRE (The elimination and choice translating reality)	A literature review on sustainable energy planning existing shows uncountable applicability of MCDM methods in socio-economic scenario
Karaca et al. [14]	Solar Energy Potential of Konya and Its Surroundings and Application of Solar Electricity Production System in Selçuklu Municipality Headman's Houses	The solar energy potential of Konya was calculated and solar panel application was made in the headman's houses	Analysis	It constitutes an example of renewable energy use for the district municipality
Dogrusoy and Serin [22]	Architectural Examination of the Potential of Renewable Energy Resources in the City of Izmir	İzmir districts were evaluated in terms of energy potential using one of the MCDM techniques	MCDM AHP	It sets an example for the use of MCDM on a district basis
Adam et al. [12]	Methodologies for city-scale assessment of renewable energy generation potential to inform strategic energy infrastructure investment	Energy investment decision has been made for Leeds city council.	Mapping	It is an example of energy use in international municipalities

(continued)

Table 1. (*continued*)

Author	Article	Purpose	MCDM method	Contribution
Arikan and Cam [18]	Web Based Feasibility Analysis of Wind and Solar Energy Systems	Amasrada made a potential analysis of solar and wind energies	Rayleigh statistical method	It constitutes an example of decision making on a district basis
Guven [15]	Meeting the Energy Need of Bahçelievler Municipality Building with Solar and Wind System, Optimization and Cost Analysis	Energy source decision has been made for Bahçelievler municipality building	Analysis	It constitutes an example of renewable energy use for the district municipality
Kumar et al. (2017)	A review of multi criteria decision making (MCDM) towards sustainable renewable energy development	Comparisons were made between MCDM techniques	Analysis	In this study several MCDM methods are introduced and the possibility of usage is searched to find the future optimal renewable energy application procedure
Biberci et al. [16]	Local Alternative Renewable Energy Sources Use in Administration and Financial Returns	An investigation has been made on the use of renewable energy in Turkey's local governments	Analysis	It shows the lack of use of renewable energy
Gazhelia and Berghb [17]	Real options analysis of investment in solar vs. wind energy: Diversification strategies under uncertain prices and costs	It has decided to invest between solar and wind energies	Real Option Analysis	It constitutes an example of investment decision between two energy sources

(*continued*)

Table 1. (*continued*)

Author	Article	Purpose	MCDM method	Contribution
Topcui et al. [7]	The evaluation of electricity generation resources: The case of Turkey	Turkey's energy resources have been compared and evaluated in nuber of ways	MADM (Multiple-attribute decision making)	It affords robust recommendations for energy policymakers in Turkey by taking into account by several dimensions. It is seen that Turkey has to growth the share of original and renewable electricity generation resources
Emeksiz and Demirci [9]	The determination of offshore wind energy potential of Turkey by using novelty hybrid site selection method	The wind potential of the coastal regions of Turkey has been examined	MCDM	The methodology suggested in this study can be used to attain a decision making processat the regional level for offshore wind farm planning
Argin et al. [10]	Exploring the offshore wind energy potential of Turkey based on multicriteria site selection	The wind potential of the coastal regions of Turkey has been examined	Mapping	The proposed methodology can be applied to make a decision to solve a problem about offshore wind energy plant in the regional base
Taminiau and Byrne [13]	City-scale urban sustainability: Spatiotemporal mapping of distributed solar power for New York City	The solar energy potential of New York City was evaluated	Geographic information system (GIS)	The interrelations between the production and consumption across the energy networks are not searched in this study since more detailed works are required to see the opportunities for development of solar cities

(*continued*)

Table 1. (*continued*)

Author	Article	Purpose	MCDM method	Contribution
Li et al. [20]	The sustainable development-oriented development and utilization of renewable energy industry-A comprehensive analysis of MCDM methods	An energy investment analysis has been made for China	MCDM	It presents its application by comparing different techniques of MCDM
Yilan et al. [19]	Analysis of electricity generation options for sustainable energy decision making: The case of Turkey	Turkey has been evaluated in terms of energy potentials	MCDM	It shows the energy potential of Turkey
Celik and Ozgur [8]	Review of Turkey's photovoltaic energy status: Legal structure, existing installed power and comparative analysis	The development of Turkey's PV use was examined and compared with 5 European countries	Analysis	It shows the situation of Turkey in terms of solar energy according to European standards

3 Methodology

Although renewable energy is preferred more due to the increasing energy demand and the limited use of fossil fuels, it has disadvantages due to intermittency and low efficiency. In addition, the dependence of renewable energy sources on climatic and geographical conditions makes the source from which the highest efficiency can be obtained different for each region.

In this case, it is necessary to apply optimization methods in order to make the most proper choices (source, location, size) according to the conditions of the region where the power plant will be established. This approach has shown a geometric increase over the last 20 years [27]. The reason for this is the development of optimization techniques and computer technology over time, and problems that were once considered impossible to solve can be easily solved by computer algorithms.

Since switching to a new renewable energy source instead of fossil fuels in current use will cause radical change in the long run; It is a situation that affects many areas, both economic, environmental and social. Various approaches have emerged to take the right decision. While methods such as linear programming and mix-integer are used for problems such as region determination and system capacity increase [29], genetic algorithms are preferred for problems such as calculating plant dimensions that will optimize cost and emissions [29].

MCDM is the mostly used method to decide on the source to be used. As mentioned earlier, making this decision will affect many different areas; therefore, a multidimensional assessment is needed.

MCDM is a method that deals with the decision-making process between multiple conflicting objectives. This method needs a decision maker to decide. There are many different MCDM techniques and the most suitable one among these techniques should be chosen in order to find an optimum solution [21].

MCDM techniques have a wide area of use outside of energy systems. There are usage examples in many fields such as agriculture, industry, and economy. Examples include the selection and layout of storage areas [30], maritime operations and maritime logistics [31], placing distribution centers of some warehouses in the best possible locations in the military logistics system [32], portfolio optimization related to risk and return [33].

At this point, it is useful to mention MCDM techniques. It is not right to distinguish between good and bad for any technique because each technique has its own advantages and disadvantages [20]. Among the techniques those are used to decide on the energy source, it is clear that AHP is the widely used technique among all. [21] and [34] reached conclusions that confirm this.

However, it cannot be denied the fact that in recent years, the fuzzy extension of AHP became more and more popular. Although this technique has been known since Zadeh's proposal at 1965 [34] and acceleration on devolopment is recorded after Chang's [35] study at 1996. As explained in the study of Liu et al. [36], Fuzzy AHP has a wide range of uses. This includes the energy sector. It has been supported by many studies in the literature that it is very suitable for decision-making problems between energy sources, as in the study of Ligus and Peternek [37].

Since this study also includes an energy source decision, it was decided to determine the decision by using one of the MCDM methods, Fuzzy AHP. This method is one of the most common MCDM techniques in the literature and it is examined the usage examples in similar studies in Literature Survey.

3.1   Fuzzy AHP

The AHP is one of the many MCDM methods. By using AHP, criteria weight calculation can be made to be used in other weighted decision making methods and calculations can be made to decide between these criteria. AHP was first developed by Thomas L. Saaty in the 1970s and is a method that enables the decision maker to make a decision by placing a complex problem in a hierarchy [38].

The pairwise comparison is performed by employing a ratio scale in the conventional AHP. To emphasize the decision or preference a nine-point scale is used between choices those may be defined as equally, moderately, strongly, very strongly, or extremely preferred. This nine-point discrete scale is simple and easy to use, but it does not evaluate the uncertainty coming from the identifying the choices as a number. It is not disregarded subjectivity and imprecision of human decisions on relative importance of criteria. The verbal expressions are always related to the sensations and are not precise every time

[39]. Instead of defining verbal expressions and representing the importance level of criteria, it is easy to represent them by using objective, incalculable and precise numbers. These problems can be converted a fuzzy model, if its boundaries became unclear. Because of this reason, the Saaty scale is replaced Triangular Fuzzy Numbers for the pair wise comparison part [40].

In the AHP method, the purpose is determined first. The top rung of the hierarchy is purpose. The next step is criteria and sub-criteria; the last step is definition and evaluation of alternatives [41].

Please note that the first paragraph of a section or subsection is not indented. The first paragraphs that follows a table, figure, equation etc. does not have an indent, either.

$$A = \begin{bmatrix} 1 & a_{12} & \cdots & a_{1n} \\ a_{21} & 1 & \cdots & a_{21} \\ \vdots & \vdots & \ddots & \vdots \\ a_{n1} & a_{n2} & \cdots & 1 \end{bmatrix} \tag{1}$$

Each a_{ij} term in this matrix represents the importance value of criterion i relative to criterion j. This significance value is decided according to Table 2.

Table 2. AHP Saaty's scale of importance and corresponding triangular fuzzy numbers [41]

Importance value	Meaning	Triangular fuzzy numbers
1	Equally important	1,1,1
2	Intermediate value between 1 and 3	1,2,3
3	Slightly important	2,3,4
4	Intermediate value between 3 and 5	3,4,5
5	Important	4,5,6
6	Intermediate value between 5 and 7	5,6,7
7	Strongly important	6,7,8
8	Intermediate value between 7 and 9	7,8,9
9	Strongly important	9,9,9

$\mu_M(x): R \to [0, 1]$ is equal to the membership function:

$$x = \begin{cases} \frac{x}{m-l} - \frac{l}{m-l}, & x \in [l, m], \\ \frac{x}{m-u} - \frac{u}{m-u}, & x \in [m, u] \\ 0, & otherwise \end{cases} \tag{2}$$

where $l \leq m \leq u$, l and u denote the lowest and highest values of the support of M, respectively, and m is the mid-value of M. When $l = m = u$, it is a non-fuzzy number traditionally. The triangular numbers M_1, M_3, M_5, M_7, M_9 re used to show the choice from "equal to extremely preferred," and $M_2, M_4, M_6,$ and M_8 are interim values.

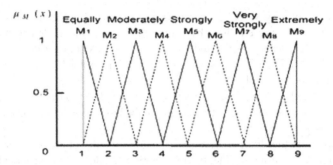

Fig. 1. The membership functions of the triangular numbers [39]

Figure 1 shows the triangular fuzzy numbers $M_t = (lt, mt, ut)$ where $t = 1, 2... 9$ and where lt and ut are the lower and upper values of the fuzzy number Mt respectively [39]:

For the consistency check, first need to find the λ_max eigenvector. After λ_max is found, the following operations are performed in order to obtain the consistency ratio;

$$CI = \frac{maks - n}{n - 1} \tag{3}$$

$$CR = \frac{CI}{RI} \tag{4}$$

In this equation, RI is a constant determined by n. The RI constant to be used according to the number n is given in Table 3 [41].

Table 3. RI constants for the AHP consistency check [39].

N	1	2	3	4	5	6	7	8	9	10
RI	0	0	0,58	0,9	1,12	1,24	1,32	1,41	1,46	1,49

Then to calculate the weights, the following equation is applied for each criterion;

$$\tilde{r} = \left(\prod_{j=1}^{n} \tilde{d}_{ij} \right)^{1/n}, i = 1, 2, \ldots \tag{5}$$

Here d is the average triangular fuzzy number and n stands for the size of the matrix. After this process, $\tilde{w}_i$ is found by [33];

$$\tilde{w}_i = \tilde{r}_1 \times \left(\sum_{j=1}^{n} \tilde{r}_j \right)^{-1} = (lw_i, mlw_i, ulw_i) \tag{6}$$

Finally, the relative non fuzzy weight is calculated for each criterion [42];

$$M_i = \frac{lw_i + mlw_i + ulw_i}{3} \tag{7}$$

M_i is normalized;

$$M_{i_n} = \frac{M_i}{\sum_{i=1}^{n} M_i} \tag{8}$$

4   Model Application and Results

To implement the Fuzzy AHP method at first a hierarchy diagram must be constructed. In this diagram, the target is specified in the first layer and criteria are determined in the second layer in order to reach this target. There may be as many sub-criteria as necessary from the main criteria. In the last layer, the alternatives are determined and the diagram is completed. Then, pairwise comparisons of these alternatives and criteria will be made in the mathematical model [42].

In the proposed model, the goal is to determine the renewable energy source to be invested for the province of Balıkesir. Criteria for this target have been determined as technical, economic, environmental and social. There are sub-criteria for each main criterion. Some judgements will be prepared between 3 alternative energy sources, namely solar, geothermal and wind (Fig. 2).

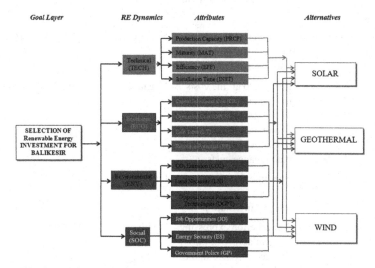

Fig. 2. A hierarchy of renewable energy source alternatives for Balıkesir

4.1   Criteria

When the similar studies in the literature are evaluated to determine the criteria, it is seen that 4 main criteria are used in previous studies [19] and [43] mentioned earlier in this article. As it is mentioned before, these criteria are technical, economic, environmental and social criteria. Stein's study [44] also chose the same main criteria, supporting this decision. The explanation of the sub-criteria for each criterion is given below.

Table 4. Importance weights for the fuzzy AHP model

Criteria	Description	Benefit attribute	Reference
Technical			
Production capacity	Corresponds to the amount of energy to be produced by the power plant to be established	Positive	[20, 45]
Maturity	It expresses the recognition of technology and the awareness of its weak points	Positive	[20, 35, 43, 45–47]
Efficiency	Refers to the technology's percentage of efficiency	Positive	[19, 20, 33, 43, 45–47]
Installation time	Time: It is the criterion that measures how long the plant can be installed	Negative	[45]
Economic			
Capital investment	The capital required for the establishment of the power plant	Negative	[20, 35, 43, 45–47]
Operation cost	It refers to all the costs to be incurred each year to maintain the power plant (such as the energy it spends to operate, maintenance costs..)	Negative	[20, 35, 43, 45–47]
Life time	The time that the plant can spend in operation after its installation	Positive	[35, 46]
Resource potential	The criterion expressing the availability of the selected resource in the region	Positive	[20, 35]
Environmental			
CO_2 emission	It refers to the carbon emission that will occur during production	Negative	[19, 20, 35, 45–47]

(*continued*)

Table 4. (*continued*)

Criteria	Description	Benefit attribute	Reference
Land necessity	The size of the land needed for the installation of the power plant	Negative	[19, 20, 35, 43, 46, 47]
Disposal green politics and technologies	It refers what kind of policies are followed on the recycling of the wastes to be generated and what level of technology is available in this regard	Positive	[47–50]
Social			
Job opportunities	The employment rate to be created by the power plant to be established	Positive	[19, 20, 35, 43, 46, 47]
Energy security	It refers to the degree of continuity that energy can be obtained	Positive	[47, 51–53]
Goverment policy	It is the criterion that measures how supportive the policies of the state towards the resource to be selected are	Positive	[20, 47]

4.2   Evaluation Weights of Individual Levels of Proposed Hierarchy

The succeeding charge of the decision makers is the comparison of the elements for a defined level of a pairwise base after construction the hierarchy, to foresee the relation between the relative importance and the proceeded element of the model. The Renewable Energy Dynamics of proposed AHP model is evaluated by three experts in energy. These experts are elected from the pool of the persons who are experienced about the region and the energy. After their evaluation, the mean of pairwise comparison results is calculated. The weights of criteria at each level are determined by using these results of calculations and the triangular fuzzification values seen at Table 2. Moreover, the desired choice is the one having the highest weight between the alternatives. The normalized energy dynamics and their characteristics can be obtained by Fuzzy AHP as shown at Table. 5.

The aggregated result for each alternative according to the normalized weight of each criterion is seen in Table 6.

After computing the CR of all pairwise comparison matrices, it was seen that all of them are less than 10% [37]. Consequently, it can be said that the consistency of the decision in all the pairwise comparison matrices is acceptable.

Table 5. Importance weights for the fuzzy AHP model

RE dynamics	Normalized weights	Attributes	Normalized weights
Technical	0,37	Efficiency	0,062
		Production capacity	0,051
		Installation time	0,069
		Maturity	0,068
Economic	0,29	Capital investment	*0,116*
		Operation cost	0,033
		Lifetime	0,051
		Resource potential	0,049
Social	0,35	Job opportunities	*0,103*
		Energy security	0,069
		Government policies	0,079
Environmental	0,37	CO_2 emission	*0,103*
		Disposal green policies and technologies	*0,113*
		Land necessity	0,033

Table 6. Aggregated results for each alternative according to each attribute

	Attribute weights	Weights	Solar	Geothermal	Wind
Efficiency	0,246	0,062	0,461	0,227	0,312
Production capacity	0,205	0,051	0,358	0,507	0,135
Installation time	0,276	0,069	0,521	0,300	0,179
Maturity	0,273	0,068	0,442	0,442	0,116
Capital investment	0,463	0,116	0,294	**0,541**	0,165
Operation cost	0,133	0,033	0,306	**0,495**	0,199
Lifetime	0,206	0,051	**0,495**	0,199	0,306
Resource potential	0,198	0,049	0,315	0,411	0,275
Job opportunities	0,411	0,103	0,315	0,411	0,275
Energy security	0,275	0,069	**0,589**	0,239	0,178
Government policies	0,315	0,079	0,306	0,495	0,199
CO_2 emission	0,414	0,103	0,294	**0,541**	0,165
Disposal green policies and technologies	0,453	0,113	0,306	0,495	0,199
Land necessity	0,133	0,033	**0,729**	0,186	0,085
Total	4,000	1,000	**0,384**	**0,416**	**0,200**

5 Conclusions and Future Work

Solar, wind, hydro, and geothermal heat are the natural sources to generate the renewable energy. It is seen that the renewable energy usage is making important contributions by supplying the energy requirements and offering chance for new establishments and employments to the economies of the developing countries. Turkey is also a developing country but the depending on the energy import to supply the requirement. The most energy consuming and one of the most populated city of north-west region is Balikesir. In the future sight, it is seen that Balikesir will be keep its position in energy consumption. Because of this reason, this study is focused on the energy demand and supply of Balikesir. The selection between the renewable energy sources is performed by using the fuzzy AHP technique.

Determining the relative importance of renewable energy resource alternatives is a fundamental problem in selection for local authorities. The AHP supplies the required abilities for solving this problem. Besides the disadvantage of using one to nine discrete scale in conventional AHP method, the uncertainty and conflicts of the resources alternatives are not evaluated. The determination of relative importance of resources preferences is depend on the subjective sight and judgements. Here, in this study the importance weighting for alternative renewable energy sources is identified by a fuzzy AHP having an extend analysis which is an effective technique since it can overcome the conflicts in human decisions. The fuzzy AHP algorithm having extend analysis simplifies the establishment of weight vectors then application of them will be easier by eliminating the depressing calculation of eigenvectors necessary in the conventional AHP. In this paper, to exhibit of method, a renewable energy source selection problem is modelled.

When technical dynamics is considered in this model, the ranking of attributes is similar with all other criterion. When Economic dynamics is reflected, the ranking order of attributes in descending order as capital investment, lifetime, resource potential and operation cost. Capital investment is observed as the most dominant. When Social dynamics is measured, the job opportunities and government policies has higher priority than energy security attribute. When Environmental dynamics is considered, the ranking of attributes is the similar as all other attributes.

The optimal renewable energy resource for the location under consideration is selected as the geothermal energy according to the MCDM analysis results. The alternative resources are sorted in the descending order are geothermal, solar, and wind. The geothermal energy resource to generate electricity, is detected as important from the point of investment. The solar resource can be evaluated as another investment alternative since it has just close weight with the geothermal. The result of this study is consistent with actual situation, i.e., solar and geothermal resources are being preferred in Turkey since they are more effective and clean for environment discussed by Hepbasli and Utlu [54].

As explained by Dursun and Alboyacı, the power density, average speed and the capacity of wind have the most preferable values in some regions of Balikesir that has the best conditions to generate electricity with the wind tribunes [25]. However, in this proposed model, it is observed that the model followed a holistic approach because the dynamics weights, which are considered technical, economic, social and environmental, is very close.

This study is limited since the evaluated criteria in the prioritization analysis are restricted. If required, these evaluations can be performed by developing the further criteria in the future. Numerous MCDM methods as Fuzzy VIKOR, fuzzy PROMETHEE, fuzzy ELECTRE, or fuzzy TOPSIS can be employed in the future studies to comparing their outputs. The techniques of establishing the correct fuzzy numbers can be explained in the further studies.

References

1. Khalil, H.B., Zaidi, S.J.: Energy crisis and potential of solar energy in Pakistan. Renew. Sustain. Energy Rev. **1**(31), 194–201 (2014)
2. Mertoglu, O., Simsek, S., Basarir, N., Paksoy, H.: Geothermal energy use, country update for Turkey. In: Proceedings of the European Geothermal Congress, pp. 11–14, Den Haag, The Netherlands (2019)
3. Kaygusuz, A.: Geothermal energy for clean and sustainable development in Turkey. J. Eng. Res. Appl. Sci. **8**(1), 1041–1050 (2019)
4. Lakatos, L., Hevessy, G., Kovács, J.: Advantages and disadvantages of solar energy and wind-power utilization. World Fut. **67**(6), 395–408 (2011)
5. Brophy, P.: Environmental advantages to the utilization of geothermal energy. Renew. Energy **10**(2–3), 367–377 (1997)
6. Dincer, I., Ozturk, M.: Geothermal Energy Systems, pp. 57–83. Elsevier, Oxford (2021)
7. Topcu, I., Ülengin, F., Kabak, Ö., Isik, M., Unver, B., Ekici, S.O.: The evaluation of electricity generation resources: the case of Turkey. Energy **167**, 417–427 (2019)
8. Celik, A.N., Özgür, E.: Review of Turkey's photovoltaic energy status: legal structure, existing installed power and comparative analysis. Renew. Sustain. Energy Rev. **134**, 110344 (2020)
9. Emeksiz, C., Demirci, B.: The determination of offshore wind energy potential of Turkey by using novelty hybrid site selection method. Sustain. Energy Technol. Assess. **36**, 100562 (2019)
10. Argin, M., Yerci, V., Erdogan, N., Kucuksari, S., Cali, U.: Exploring the offshore wind energy potential of Turkey based on multi-criteria site selection. Energy Strat. Rev. **23**, 33–46 (2019)
11. Melikoglu, M.: Geothermal energy in Turkey and around the world: a review of the literature and an analysis based on Turkey's vision 2023 energy targets. Renew. Sustain. Energy Rev. **76**, 485–492 (2017)
12. Adam, K., Hoolohan, V., Gooding, J., Knowland, T., Bale, C.S., Tomlin, A.S.: Methodologies for city-scale assessment of renewable energy generation potential to inform strategic energy infrastructure investment. Cities **54**, 45–56 (2016)
13. Taminiau, J., Byrne, J.: City-scale urban sustainability: spatiotemporal mapping of distributed solar power for New York City. Wiley Interdiscipl. Rev. Energy Environ. **9**(5), e374 (2020)
14. Karaca, İ.H., Gürkan, E.C., Yaparh, H.: Konya ve Civarının Güneş Enerjisi Potansiyeli ve Selçuklu Belediyesi Muhtar Evlerinde Güneşten Elektrik Üretim Sistemi Uygulaması. I Konya Kent Sempozyumu, pp. 275–292 (2011)
15. Güven, A.F.: Bahçelievler Belediye Başkanlık Binasının Enerji İhtiyacının Güneş ve Rüzgar Sistemi ile Karşılanması, Optimizasyonu ve Maliyet Analizi. Sinop Üniversitesi Fen Bilimleri Dergisi. **2**(1), 24–36 (2016)
16. Biberci, M., Doğan, M., Dilber, C., Çelik, M.: Alternatif Yenilenebilir Enerji Kaynaklarının Yerel Yönetimlerde Kullanımı Ve Mali Getirileri, Belediyelerin Geleceği ve Yeni Yaklaşımlar, pp. 541–560 (2017)

17. Gazheli, A., van den Bergh, J.: Real options analysis of investment in solar vs. wind energy: diversification strategies under uncertain prices and costs. Renew. Sustain. Energy Rev. **82**, 2693–2704 (2018)
18. Arikan, Y., Ertuğrul, Ç.A.: Implementation of feasibility analysis of wind and solar energy on the web base. Int. J. Eng. Res. Dev. **9**(1), 1–10 (2017)
19. Yilan, G., Kadirgan, M.N., Çiftçioğlu, G.A.: Analysis of electricity generation options for sustainable energy decision making: the case of Turkey. Renew. Energy **146**, 519–529 (2020)
20. Li, T, Li, A., Guo, X.: The sustainable development-oriented development and utilization of renewable energy industry——a comprehensive analysis of MCDM methods. Energy **212**, 118694 (2020)
21. Pohekar, S.D., Ramachandran, M. Application of multi-criteria decision making to sustainable energy planning—a review. Renew. Sustain. Energy Rev. **8**(4), 365–381 (2004)
22. Türkseven Dogrusoy, İ., Serin, E.: İzmir kentindeki yenilenebilir enerji kaynaklarının potansiyelinin mimari açıdan irdelenmesi [Analysis of the potencials of renewable energy sources in Izmir city in architectural point of view]. Dokuz Eylül Üniversitesi Mühendislik Fakültesi Mühendislik Bilimleri Dergisi **15**(3), 1–25 (2013)
23. Karatop, B., Taşkan, B., Adar, E., Kubat, C.: Decision analysis related to the renewable energy investments in Turkey based on a fuzzy AHP-EDAS-fuzzy FMEA approach. Comput. Ind. Eng. **151**, 106958 (2021)
24. Wang, Y., Xu, L., Solangi, Y.A.: Strategic renewable energy resources selection for Pakistan: based on SWOT-fuzzy AHP approach. Sustain. Cities Soc. **52**, 101861 (2020)
25. Dursun, B., Alboyaci, B.: An evaluation of wind energy characteristics for four different locations in Balikesir. Energy Sour Part A **33**(11), 1086–1103 (2011)
26. Ilten, N., Selici, A.T.: Investigating the impacts of some meteorological parameters on air pollution in Balikesir. Turkey. Environ Monit. Assess **140**(1), 267–277 (2008)
27. Banos, R., Manzano-Agugliaro, F., Montoya, F.G., Gil, C., Alcayde, A., Gómez, J.: Optimization methods applied to renewable and sustainable energy: a review. Renew. Sustain. Energy Rev. **15**(4), 1753–1766 (2011)
28. Cai, Y.P., Huang, G.H., Yang, Z.F., Lin, Q.G., Tan, Q.: Community-scale renewable energy systems planning under uncertainty—an interval chance-constrained programming approach. Renew. Sustain. Energy Rev. **13**(4), 721–735 (2009)
29. Soroudi, A., Ehsan, M., Zareipour, H.: A practical eco-environmental distribution network planning model including fuel cells and non-renewable distributed energy resources. Renew. Energy **36**(1), 179–188 (2011)
30. Chang, N.B., Parvathinathan, G., Breeden, J.B.: Combining GIS with fuzzy multicriteria decision-making for landfill siting in a fast-growing urban region. J. Environ. Manage. **87**(1), 139–153 (2008)
31. Chou, C.C.: A fuzzy MCDM method for solving marine transshipment container port selection problems. Appl. Math. Comput. **186**(1), 435–444 (2007)
32. Farahani, R.Z., Asgari, N.: Combination of MCDM and covering techniques in a hierarchical model for facility location: a case study. Eur. J. Oper. Res. **176**(3), 1839–1858 (2007)
33. Ehrgott, M., Klamroth, K., Schwehm, C.: An MCDM approach to portfolio optimization. Eur. J. Oper. Res. **155**(3), 752–770 (2004)
34. Zadeh, L.A.: Fuzzy sets. In: Zadeh, L.A. (ed.) Fuzzy sets, fuzzy logic, and fuzzy systems: selected papers, pp. 394–432 (1996)
35. Wang, J.J., Jing, Y.Y., Zhang, C.F., Zhao, J.H.: Review on multi-criteria decision analysis aid in sustainable energy decision-making. Renew. Sustain. Energy Rev. **13**(9), 2263–2278 (2009)
36. Liu, Y., Eckert, C.M., Earl, C.: A review of fuzzy AHP methods for decision-making with subjective judgements. Exp. Syst. Appl. **161**, 113738 (2020)

37. Ligus, M., Peternek, P.: Determination of most suitable low-emission energy technologies development in Poland using integrated fuzzy AHP-TOPSIS method. Energy Procedia **153**, 101–106 (2018)
38. Saaty, T.L.: Fundamentals of Decision Making and Priority Theory with the Analytic Hierarchy Process. RWS Publication, Pittsburg (2000)
39. Kwong, C.K., Bai, H.: Determining the importance weights for the customer requirements in QFD using a fuzzy AHP with an extent analysis approach. IIE Trans. **35**(7), 619–626 (2003)
40. Moayeri, M., Shahvarani, A., Behzadi, M.H.: The application of fuzzy analytic hierarchy process in high school math teachers ranking. Math. Educ. Trends Res. **1**, 20–30 (2016)
41. Kaganski, S., Majak, J., Karjust, K.: Fuzzy AHP as a tool for prioritization of key performance indicators. Procedia Cirp **72**, 1227–1232 (2018)
42. Soberi, M.S., Ahmad, R.: Application of fuzzy AHP for setup reduction in manufacturing industry. J. Eng. Res. Educ **8**, 73–84 (2016)
43. Lee, H.C., Chang, C.T.: Comparative analysis of MCDM methods for ranking renewable energy sources in Taiwan. Renew. Sustain. Energy Rev. **92**, 883–896 (2018)
44. Stein, E.W.: A comprehensive multi-criteria model to rank electric energy production technologies. Renew. Sustain. Energy Rev. **22**, 640–654 (2013)
45. Cavallaro, F., Ciraolo, L.: A multicriteria approach to evaluate wind energy plants on an Italian island. Energy Policy **33**(2), 235–244 (2005)
46. Şengül, Ü., Eren, M., Shiraz, S.E., Gezder, V., Şengül, A.B.: Fuzzy TOPSIS method for ranking renewable energy supply systems in Turkey. Renew. Energy **75**, 617–625 (2015)
47. Alipour, M., Alighaleh, S., Hafezi, R., Omranievardi, M.: A new hybrid decision framework for prioritizing funding allocation to Iran's energy sector. Energy **121**, 388–402 (2017)
48. Coyle, G.: The not-so-green renewable energy: Preventing waste disposal of solar photovoltaic (PV) panels. Golden Gate U. Envtl. LJ. **4**, 329 (2010)
49. Recycling revolution necessary to complete the clean energy transition (2021). https://www.cleanenergywire.org/news/recycling-revolution-necessary-complete-clean-energy-transition. Accessed 18 June 2021
50. EPA Releases Briefing Paper on Renewable Energy Waste Management (2021) https://www.epa.gov/newsreleases/epa-releases-briefing-paper-renewable-energy-waste-management. Accessed 18 June 2021
51. Augutis, J., Martišauskas, L., Krikštolaitis, R.: Energy mix optimization from an energy security perspective. Energy Convers. Manage. **90**, 300–314 (2015)
52. Francés, G.E., Marín-Quemada, J.M., González, E.S.: RES and risk: renewable energy's contribution to energy security. a portfolio-based approach. Renew. Sustain. Energy Rev. **26**, 549–559 (2013)
53. Johansson, B.: Security aspects of future renewable energy systems–a short overview. Energy **61**, 598–605 (2013)
54. Hepbasli, A., Utlu, Z.: Evaluating the energy utilization efficiency of Turkey's renewable energy sources during 2001. Renew. Sustain. Energy Rev. **8**(3), 237–255 (2004)

Author Index

Printed in the United States
by Baker & Taylor Publisher Services